Environment and Welfare

Environment and Welfare

Towards a Green Social Policy

Edited by

Tony Fitzpatrick
Senior Lecturer in the School of Sociology and Social Policy
University of Nottingham

and

Michael Cahill
Reader in the Department of Applied Social Studies
University of Brighton

First published 2002 by
PALGRAVE MACMILLAN
Houndmills, Basingstoke, Hampshire RG21 6XS and
175 Fifth Avenue, New York, N.Y. 10010
Companies and representatives throughout the world

PALGRAVE MACMILLAN is the global academic imprint of the Palgrave
Macmillan division of St. Martin's Press, LLC and of Palgrave Macmillan Ltd.
Macmillan® is a registered trademark in the United States, United Kingdom
and other countries. Palgrave is a registered trademark in the European
Union and other countries.

ISBN 0–333–91984–X

This book is printed on paper suitable for recycling and made from fully
managed and sustained forest sources.

A catalogue record for this book is available from the British Library.

Library of Congress Cataloging-in-Publication Data

Environment and welfare: towards a green social policy/editors, Tony
Fitzpatrick, Michael Cahill.
        p. cm.
    Includes bibliographical references and index.
    ISBN 0–333–91984–X
        1. Social policy – Environmental aspects. 2. Environmental policy.
    I. Fitzpatrick, Tony, 1966–II. Cahill, Michael, 1951–

    HN17.5 .E58 2002
    361.6'1–dc21

                                                    2002072310

10   9   8   7   6   5   4   3   2
11   10   09   08   07   06   05   04   03

Printed and bound in Great Britain by
Antony Rowe Ltd, Chippenham, Wiltshire

Contents

List of Tables

List of Figures

List of Abbreviations

BI	Basic Income
CHP	Combined Heat and Power Systems
CRI	Capital Receipts Initiative
DETR	Department for the Environment, Transport and the Regions
DSS	Department of Social Security
DTI	Department of Trade and Industry
EESOP	Energy Efficiency Standards of Performance Scheme
ENDS	Environmental Data Services
EST	Energy Savings Trust
GDP	Gross Domestic Product
GNP	Gross National Product
HECA	Home Energy Conservation Act
HEES	Home Energy Efficiency Scheme
ISEW	Index of Sustainable Economic Welfare
LETS	Local Exchange and Trading Systems
MEW	Measure of Economic Welfare
NCC	National Consumer Council
NDP	Net Domestic Product
NGO	Non-Governmental Organisation
OFWAT	Office of Water Services
RCEP	Royal Commission on Environmental Pollution

Notes on the Contributors

John Barry is Reader in the School of Politics, Queen's University, Belfast. His main publications are *Rethinking Green Politics* (1999), *Environment and Social Theory* (1999) and the *International Encyclopedia of Environmental Politics* (2001), co-edited with Gene Frankland.

Michael Cahill is Reader in the Department of Applied Social Studies, University of Brighton. His main publications are *The New Social Policy* (1994), *The Environment and Social Policy* (2001) and *Environmental Issues and Social Welfare* (2001), co-edited with Tony Fitzpatrick.

Tony Fitzpatrick is Senior Lecturer in the School of Sociology and Social Policy, University of Nottingham. His main publications are *Freedom and Security* (1999), *Welfare Theory* (2001) and *Environmental Issues and Social Welfare* (2001), co-edited with Michael Cahill.

Meg Huby is Senior Lecturer in the Department of Social Policy and Social Work, University of York. Her publications include *Fair and Sustainable* (1997), with B. Hills and P. Kenway, *Social Policy and the Environment* (1998) and *A Study of Town Life* (1999) with J. Bradshaw and A. Corden.

Mathew Humphrey is Lecturer in Political Theory at the Department of Politics, University of Nottingham. He is editor of *Political Theory and the Environment: A Reassessment* (2001) and is currently completing a manuscript on the political philosophy of nature preservation.

Tim Jackson is Professor in the Centre for Environmental Strategy, University of Surrey. His publications include *Material Concerns* (1996), *Power in Balance* (1997) and *Sustainable Economic Welfare in the UK* (1997), with N. Marks, J. Ralls and S. Stymne.

Adrian Little is Lecturer in the Department of Social Policy and Politics, Goldsmiths College, University of London. His main publications include *The Political Thought of André Gorz* (1996), *Post-Industrial Socialism* (1998) and *The Politics of Community* (2002).

James Robertson is an independent writer and lecturer, formerly an official in the Cabinet Office and a director of inter-bank research. His recent publications include *Beyond the Dependency Culture* (1997),

Transforming Economic Life (1998), *The New Economics of Sustainable Development* (1999) and *Creating New Money* (2000), with Joseph Huber.

Colin C. Williams is Reader in Economic Geography in Department of Geography, University of Leicester. His publications include *Informal Employment in the Advanced Economies* (1998), *A Helping Hand* (1999) and *Revitalising Deprived Urban Neighbourhoods* (2001), all with Jan Windebank.

1
The New Environment of Welfare

Tony Fitzpatrick and Michael Cahill

Introduction: what is sustainability?

Green ideas began to influence the study of social policies in the late 1980s (Cahill, 1991; Ferris, 1993; Pierson, 1998), swiftly becoming a reference point for those who were unhappy with the radical Right but who also felt that something other than a return to the traditional welfare state was required. This influence has left an ambiguous legacy, however. Rather like a song which is as popular as it is forgettable, the rise of environmentalism has been a little too easy, a little too effortless. Although the introductory textbooks which fail to mention it are now rare (George and Wilding, 1994; Barry, 1998; Cahill, 1999) this reception is as reflective of the discipline's tendency to assimilate new ideas into a taken-for-granted framework as it is of ecologism's actual significance.[1] Environmentalism risks becoming just another appendix to the usual array of welfare ideologies. Nevertheless, the sustained work on environmentalism and social policy which has emerged in recent years (Fitzpatrick, 1998; Huby, 1998; Trainer, 1998; Cahill, 2001; Fitzpatrick with Caldwell, 2001) is beginning to suggest how and why that framework may have to change and our hope is that this book, with its companion, *Environmental Issues and Social Welfare* (Cahill and Fitzpatrick, 2001), will encapsulate the key themes and critiques that we are still only beginning to explore.

That framework is challenged by the ecological commitment to sustainability. It is always tempting to reduce theories and ideologies to one or two principles or concerns: liberalism equals liberty, communitarianism equals community, Marxism equals class, feminism equals gender, and so on. As such, environmentalism might be thought of as the social philosophy of sustainability. Yet sustainability

1

is a notoriously difficult idea to pin down. In its most famous definition sustainability implies meeting '... the needs of the present without compromising the ability of future generations to meet theirs' (Brundtland Commission, 1987: 8), yet this formulation is allied to a developmental worldview that other environmentalists have resisted. Does this contestability imply that sustainability is empty or that, on the contrary, it takes its place alongside other political concepts that are also important while being highly controversial? Our view inclines towards the latter interpretation. Those who disagree might like to pause here and offer a formulation of, say, 'equality' that is concise and uncontroversial, yet meaningful. Not easy, is it?

Essentially, environmentalists identify a disjunction between what we demand of the world and what the world is capable of supplying. If the demands we make are infinite, yet the resources upon which we can realistically draw are finite, then ours is an unsustainable existence. Sustainability therefore implies reducing human demands and/or increasing resources so that that disjunction between the two becomes less and less significant. Note that sustainability can never be an *outcome*. All life is ultimately unsustainable since all life uses energy and the law of entropy, where all energy systems run down even as the stock of energy remains constant, is inescapable in the long run. Strictly speaking, then, we have to conceive of sustainability not as a perfect ideal but as a threshold situated somewhere along a continuum of unsustainability that it is desirable for us to achieve. Sustainability is therefore a *process* where demands and resources, the infinite and finite aspects of human life, gradually conjoin.

So sustainability can be said to extend across two dimensions: space and time. Sustainability requires consistency in the use of space since there would be little point in aligning demands and resources in one country if the country next door is cheerfully polluting. It also requires consistency across time because if one generation inherits an environmentally degraded planet from its predecessors then, by definition, sustainability has not been achieved. Yet this is where the real controversies begin to kick in, of course. Can one country really trust that others are not going to defect on their environmental obligations? Can we trust that later generations will not waste the benefits that they inherit? The controversy, then, is about the distribution of benefits and burdens. Sustainability might be desirable but who is to pay for it? The picture is complicated even more when we look beyond humanity. If we could create a sustainable world by eliminating all animals should we do so? Even if we value every human life above every non-human life we might still decide that a single potential human life in the future is not worth the sacrifice of a hundred non-human lives either now or in the future.[2]

Our inclination is to say that the distribution of benefits and burdens should be in inverse proportion to the distribution of privilege. Who is to shoulder many of the burdens? The relatively advantaged. Who is to shoulder many of the benefits? The relatively disadvantaged. Sustainability therefore implies a redistribution from the affluent to the non-affluent, from present to future generations and from humans to non-humans.[3] Of course, this still leaves the question as to who are the advantaged and disadvantaged. Is the present generation of the poor more or less important than future generations of the non-poor (see Fitzpatrick, 2001a)? Furthermore, political reality always intervenes, suggesting that we should not conceive of sustainability merely as a negative sum transfer from developed to developing nations, or from the present to the future, since the consent of the present generation of the affluent is crucial and that consent is unlikely to be given if the price is too high (Fitzpatrick, 2001b). Yet even this consideration still points us in a more politically ambitious direction than we currently seem willing to travel. In short, and as a general principle, even a modest interpretation of sustainability implies a distribution across space, time and species that post-war welfare capitalism was unable or unwilling to achieve.

A similar conclusion follows when we consider more obviously ideological perspectives. How can demands and resources be made to conjoin? The first way (weak sustainability) is to expand the stock of resources. This can be done by replacing renewable resources, by substituting for non-renewable ones and by searching for technological solutions to depletion and pollution (Weizsacker *et al.*, 1998). A second way (strong sustainability) is to revise the demands that we make on the world so that, for instance, we consume far less. So, rather than adapting the world to suit ourselves we adapt ourselves to meet the finitude of nature. The third way (moderate sustainability) is to combine elements of those two approaches. Each of these implies a subtly different conception of sustainability, as explained by Andrew Dobson.

Conception A (Dobson, 1998: 41–7) is concerned to sustain the most critical aspects of natural capital, i.e. those aspects which are essential for the perpetuation of human life. It incorporates an anthropocentric rationale, in that the *needs* of future generations must override the *wants* of the present generation, and it favours three strategies for the preservation of 'critical natural capital': renewability, substitutability and conservation. Conception A is therefore one of weak sustainability since it examines ways in which we can change the natural world.

At the opposite extreme, conception C (Dobson, 1998: 50–4) identifies an intrinsic value to nature, the sustainability of which cannot therefore be measured in terms of human welfare. Of course, the former may enhance the latter but enhancing the latter cannot be the motivation for the former. Conception C therefore abandons the strategies of renewability and substitutability as violating the principles of intrinsic natural value, concentrating upon conservation as the main instrument of sustainability. It corresponds to a strong sustainability approach, where it is humans who must adapt.

Conception B (Dobson, 1998: 47–50) is a middle way, concerned with sustaining those aspects of the natural world whose loss would be irreversible. In short, while conception B acknowledges the importance of human welfare it also wants to preserve those elements of non-human nature which risk disappearing forever, regardless of the effects this might have for human welfare. It recognises the usefulness of renewability and substitutability but downgrades them in importance compared to conception A. Therefore, conception B is moderately sustainable, where both the world and humanity must be made to adapt to each other.

Now it seems clear that the policies governments have been introducing since the 1992 Rio conference fall short of even the most 'conservative' of these interpretations of sustainability, conception A. Why is this? Well, one reason is that, according to Dobson (1998: 87–164), conception A seems to require the just distribution of critical natural capital, i.e. distribution according to universal needs. This not only implies a degree of social equality that few nations have achieved but also a re-organisation of the existing pattern of ownership and the system of property rights (both based upon private accumulation), a reorganisation that social policy makers currently appear unable to contemplate. Such requirements are barely visible on the environmental agenda of developed nations, i.e. those who possess the greatest responsibility to reverse ecological decline. The preferred alternative has been 'ecological modernisation' (Hajer, 1995; Mol and Sonnenfeld, 2000) where existing institutions, practices and assumptions are adapted to meet environmental goals. As such, nations committed to welfare capitalism appear more environmentally friendly than those blinded by free markets, yet even the former still pursue levels and types of economic growth whose shelf life is declining.

In short, *the significance of ecologism for social policy lies in the belief that even modest and conservative (or weak) interpretations of sustainability challenge the traditional bases of welfare systems.* What are the principal criticisms that ecologists can make of existing welfare systems?

Main criticisms

The difficulty is answering this question lies in the fact that different schools of Green thought exist, as we have acknowledged above. To put it simply, an advocate of weak sustainability is likely to voice critiques that are considerably different to those voiced by advocates of strong sustainability. Nevertheless, if it is plausible to treat ecological thought as a distinct set of political and social ideas then there must be certain basic themes that all schools hold in common. We may debate the extent to which environmentalism is ideologically coherent and distinctive (see Chapter 3 by Mathew Humphrey) but any ideology implies a common denominator through which it marks out a theoretical field, elements of which may overlap with that of other ideologies but which is, nevertheless, constitute of itself. So, what are the common themes that inform a Green critique of social policy (cf. Fitzpatrick, 1998)?

First, and most obviously, ecologism opposes social policies and welfare systems that are unsustainable and environmentalists agree that indiscriminate economic growth is at the heart of all unsustainable practices. With consumer capitalism fuelling demands whose appetites can never be fully satisfied, pressures are placed upon the planet's ecosystem that cannot continue indefinitely. Despite this, economic growth is one of the logics of modern society whose value is virtually unquestionable. It is the economic expression of the Enlightenment vision of historical progress and social development, the means by which humanity may transcend its pre-modern origins. As such, almost all political philosophies have incorporated that logic into themselves. For Marxists, the development of productive forces was the *sine qua non* of class conflict and so of a transition to a classless society, leading most Marxists to dismiss environmentalism as a bourgeois concern (Enzensberger, 1988) although this stance has now been replaced by a more considered position (Benton, 1996). Consequently, economic growth has become a form of meta ideology, a cultural underpinning of modern thought that is too common sensical to require articulation. This meta-ideology is productivist in character, 'productivism' capturing the idea that economic growth has become an end-in-itself, *the* addiction of advanced capitalism that, like all addictions, can never be satiated.

On these grounds, social policy leaves itself open to two forms of criticism by ecologists. One criticism is that existing welfare systems are *dependent* upon productivist practices that are ultimately unsustainable. Welfare and diswelfare are usually defined and measured according to distributive patterns that trace back to the logic of productivism. The Right, for instance, treat rising material prosperity as the central justification for capitalism and for free markets in particular, the job of social policies being to assist markets by maintaining social order and enforcing the disciplines that unregulated markets require. The Left have thought it best to champion redistributive justice by emphasising a painless form of distributive politics where basic needs are met, and the less well-off assisted, by directing ever-higher levels of growth in the appropriate directions. A modest amount of direct redistribution from rich to poor may be justifiable yet, by and large, social democrats have tried to improve the share of the poorest by increasing the overall stock of national resources. Therefore, productivism can be seen to underpin all welfare regimes – a fact that possibly explains why cross-national analysis, in its search for differences, has remained oblivious to the environmental case (e.g. Esping-Andersen, 1999).

Another criticism, therefore, is that social policies *contribute* to unsustainability. Rather than redefine well-being in non-productivist directions, most welfare reforms and reform proposals have been premised on the assumption that what is wrong is not our conceptualisation of well-being but the means we have employed to realise it. With many arguing that the quality of life has not improved (and may even have declined) since the 1960s (Sennett, 1998; Putnam, 2000), and with the ascendancy of the anti-welfare Right, the general malaise of the late twentieth century has been pinned on the welfare state. Since productivism tells us that rising prosperity *must* generate more happiness and since our societies are characterised by more fear, risk consciousness, moral panic and insecurity than they have been for a long time, then the cause of this must lie with those liberal and socialist apologists for welfare institutions and public sector practices that seek to excuse away the moral defects of those who prevent us from achieving that happiness. Therefore, we need to clamp down on the work-shy, benefit fraudsters, beggars, single mums, deadbeat dads, criminals, paedophiles, etc. by ending the something-for-nothing culture and enforcing the social duties of citizenship (Garland, 2001). The draconian welfare reforms of the Right can therefore be explained

as an attempt to defend the productivist logic of free market capitalism. Yet rather than challenge this at source, many on the Left have
been persuaded to travel in similar directions. The new social
democracy (Giddens, 1998) embodies a similar attempt to distinguish
between an included majority (included by virtue of their being in paid
work or conforming to the norms of the wage contract) and an
excluded minority, the malfunctioning outsiders who represent an
ever-present danger to the middle class. So, by not re-conceptualising
well-being the defenders of the welfare state may only have undermined their own case by ceding ground to the Right.

Of course, what that re-conceptualisation implies is open to
question. For those persuaded by weak sustainability the problem may
not be economic growth *per se* but the kind of Gross Domestic Product
(GDP) growth that neglects the wide range of non-materialist indicators of well-being. For those committed to strong sustainability the
entire anthropocentric ethos of growth may need to be abandoned. Yet
for both groups it is no surprise that the welfare state's popularity
ended once it could no longer deliver the kind of well-being that productivism demands, a materialist alignment of growth and happiness.

This leads to the second Green critique of social policy, that it is too
heavily based upon wage-earning, whether as a source of material security, self-identity or social participation. The employment ethic is
essential to modernity. It was the means by which Protestantism
justified individualism and the accumulation of wealth, the means by
which industrial capitalists could sweep away the residual traces of feudalism, the means by which socialists could challenge the power of
capitalists, the means by which nation-states could compete with one
another. In facing this daunting array of political, economic and social
forces, ecologists may appear to be isolated cranks, only slightly less
dopey than flat-earthers. Doesn't everyone know that employment is
the single greatest source of meaning, purpose and prosperity for both
individuals and society as a whole?

Yet environmentalists are not quite as isolated as they first appear.
For instance, most are not arguing for the abolition of employment but
for a recognition that security, identity and participation derive from a
far wider range of sources and activities than the current orthodoxy
admits. In this respect, ecologists side with those feminists who
redefine flexibility towards employment, unpaid work and leisure away
from restrictive masculinist conceptions of economic well-being (e.g.
Reeves, 2001). Indeed, since both ecologism and feminism argue that
care, rather than productivity, is essential to well-being then the scope

for further convergence between these ideas is considerable. Furthermore, environmentalists point out that the employment ethic has been undermining itself for several decades now. With the overall decline in employment levels, though some countries are obviously more successful in this respect than others, and with the polarisation of labour markets into core and peripheral sectors, the social conditions of this ethic are fatally undermined (Gorz, 1989; Rifkin, 1995). Inevitably, the energy of the political mainstream has been expended not on revising the ethic but on blaming the usual suspects for the deterioration of those conditions, leading to welfare reforms that emphasis employment, *any* employment, as preferable to any form of non-employment. In a perverse kind of wish fulfilment it is those countries who have moved closest to the ideal of retributive workfare who have created the most jobs, though a closer analysis than their defenders seem willing to initiate reveals the cost in terms of job security, free time, family life and poverty (e.g. Gershuny, 2000).

So, rather than raising an objection that flies in the face of modernity, ecologists are trying to articulate social developments that have led modernity to undermine itself (Beck, 2000). That social policy continues to be largely insensitive to those developments is therefore a cause of some concern.

Full-time full employment, for men at least, was certainly the axis of the 'classic' welfare state. For Keynes, full employment required that an effective level of economic demand be maintained through government spending programmes; for Beveridge, welfare institutions had to be supported by a full employment economy to ensure that revenues were high enough to fund the needs of the unemployed, the sick and disabled, and the elderly. Social insurance provision was therefore ideally suited to a full employment economy; the latter enabling the former to collectivise risks, without the need for much vertical redistribution or socialised ownership to be introduced. No surprise, then, that social insurance has come under threat as fewer nations have maintained full employment levels (Clasen, 1998). In Britain and America the principle has come under attack, although insurance benefits for the relatively well-off continue to be popular in America and insurance contributions continue to be an important source of revenue in Britain (though these contributions have become more and more detached from welfare entitlements and benefit eligibility). Across Europe the principle continues to be popular but insurance funds risk falling into deficit due to higher levels of unemployment than in Britain and America and greater demographic pressures due to their more generous benefit levels.

The main alternative to a welfare system based upon social insurance is the kind of regime that has developed in Britain and America in recent years: means-tested tax credits, time-limited benefits, workfare programmes. Whether this regime be characterised as Schumpeterian (Jessop, 2001), the workfare state, the competition state, or the post-social security state (Fitzpatrick, 2001c), what is clear is that, like its predecessor, it is predicated upon the norm of wage-earning. Indeed, with demand management taken to be less effective in a globalised economy than supply-side strategies, and with millions of the unemployed and low-paid no longer trusted to participate in the labour market without incentives and/or coercion, this new regime is even more obsessed with employment than its ancestors.

In short, welfare states have developed in concert with an employment society. One school of thought supports high wages, regulation and a degree of decommodification (the right to earn insurance benefits that permit some autonomy outside the labour market); another school supports low wages, flexibility and 'enforced commodification' (the duty to be employed since the labour market is the main entrance to social inclusion). As these two schools continue to slug it out environmentalists add their voice to those calling for the redefinition of work. So that rather than 'work' being equated with 'employment' we should recognise the wider spectrum of activities that are socially valuable. This most obviously includes 'care work' but may also encompass other forms of activity that are more creative and playful, or even rebellious. If a government is trying to build a bypass that is likely to destroy the local ecosystem then who is contributing to society more: the unemployed protestors chaining themselves to trees or the underpaid security guards applying the bolt-cutters?

The difficulty comes, as before, in deciding what such environmental arguments imply for social reform. Ecological modernisers (Jacobs, 1999) are insistent that the Green case can only be promoted by stressing the job-friendly implications of Green reforms. Yet as Adrian Little suggests (Chapter 10) being job-friendly could require reductions in working-time that would not only make employment more widely available but would demand that we redefine what we mean by social participation. However, others might argue that we should try to dismantle the employment society altogether. For ecosocialists, employment implies an exploitative contract between employer and worker that is therefore less desirable than cooperative forms of worker management. For all Greens, though, the objective is to recognise that 'social wealth' has many more sources than employment, but there is an obvious problem in trying to specify the levels of contribution that non-employment activities actually make (see Chapter 6). Care work, for instance, is difficult to quantify because it has effects across time and space that are difficult to trace and, besides, commodifying it in this way seems to contradict what caring is actually about. Yet without the attempt Greens may never successfully challenge the two schools mentioned above. So, the question for environmentalists is how can we quantify that which inherently eludes quantification?

The third main critique made by environmentalists towards social policy concerns the degree of control and autonomy currently possessed by individuals. The allegation is that existing forms of social organisation and welfare provision underestimate the extent to which citizens can be self-organising. At one extreme there are libertarian and anarchistic Greens who would prefer the existence of semi-independent autarchies, or highly decentralised cooperative communities that provide for themselves with a minimal amount of external coordination (Sale, 1985). This is to envisage a Green society as that which is truly beyond state and market. The problem is that this vision may derive from pre-modern ideas of a rural, idyllic life that is premised upon communal homogeneity and the relative immobility of individuals and families. It neglects the reason why systems of national and international governance were set up in the first place: to arbitrate with impartiality between conflicting groups, clans and tribes (Locke, 1960).

Nevertheless, even if they fall short of this extreme most Greens are dissatisfied with the degree of centralisation that currently prevails. The state is interpreted largely as too distant and impersonal a set of institutions. Representative democracy is thought to encourage a passive, consumerist attitude towards the common good, one that minimises the level of political participation and organises participation around party-machines that are top-heavy (Doherty and de Geus, 1996). The assumption is that although most people care about the environment this concern has barely registered on the mainstream political agenda since the party system embodies a *status quo* that is immensely slow to respond to new developments. And even where the Greens have had some success (*Die Grunen*, for instance) they have had to compromise to such an extent that their ideological distinctiveness is eclipsed.

In terms of social policy this state centralisation is thought to encourage the 'clientalisation' of welfare, where well-being is something we receive from other sources (experts, bureaucrats) and rarely something that people collectively generate for and through themselves (Fitzpatrick, 1998). The price of paternalism and basic needs satisfaction has been an over-arching collectivism that allows little space for bottom-up provision. Greens therefore tend to support not an ethos of Victorian self-help but a new welfare settlement where the state provides a universalistic framework of regulation, accountability and basic service provision, but where greater room is made for civic associations (Hirst, 1994), decentralised policy communities (Ellison, 1999) or cooperation circles (Offe, 1996) that would control funds and allow the 'recipients' of welfare to become their own 'producers' (Barry and Proops, 2000: 93–4). In short, Green social policies seem to require a greater degree of decentralisation and the emergence of a new 'welfare citizen'.

Yet the difficulty lies not only with the state but also with a market system that is based upon the needs and demands of the most powerful economic actors, by which is meant multinational and transnational corporations. In particular, the chemical, car, nuclear and fossil fuel industries dominate the market, establishing an oligopoly that actively resists environmental legislation – even when based upon modest interpretations of sustainable development – and which lay behind the cultural and ideological assault upon ecologism in the 1990s. It is the influence of these actors upon political parties that is most worrying. By being able to buy political power they are able to undermine international cooperative agreements – symbolised by the failure of the American government to ratify the Kyoto protocols of 1997, culminating in their eventual rejection by Bush in March 2001. But Greens begin to disagree about how to rectify market centralisation. For some, the answer is more rather than less capitalism, the ideal being a Jeffersonian economy of small-scale private property and free market economics where the price mechanism is regarded as the best means of preserving environmental goods. According to ecological modernisation, the answer is to supplement market forces with a regulatory structure that actively promotes sustainability through emissions' trading, ecotaxes, recycling schemes, carbon sinks, environmental subsidies, etc. Others insist that only the collective ownership of natural resources can ultimately ensure that Green objectives are pursued by redistributing the burden and benefits of environmental sustainability (Fitzpatrick, 2001a).

These disagreements seem to replicate themselves at the level of social policy. Market reforms of the welfare state have shifted the source of well-being from the state to the commercial sector but has only enhanced the autonomy of those lucky enough (and wealthy enough) to wield power as consumers. If a new welfare citizen and Green welfare democracy is to emerge then such power has to be more widely distributed. Yet is the solution to transcend the capitalist market, by collectivising the ownership of productive property, in order to effect a new form of welfare society; to regulate market capitalism and ensure that social policies on education, health, employment and housing contribute to that regulation; or to allow capitalist markets to work more effectively by marketising all centralised forms of control? These are questions we hope this volume will inspire others to address.

These Green critiques of social policy – focusing upon growth, employment and centralisation – are not meant to be exhaustive. There are many gaps to be filled and further points to be emphasised and our hope is that *Environment and Welfare*, along with its companion, will go some way towards doing so. But before outlining the book's content and structure it is worth noting that the direction of criticism is not just one way. Social policy has potentially much to teach environmentalism.

For instance, although it has its more utopian moments the subject of social policy is a reminder that to get to where we want to be we have to start from where we are. It is very easy to construct imaginary societies that would be more desirable than our existing ones; the trick is to do so without ignoring the problem of transition and whether we can realistically expect the present to be transformed into that imaginary future. In short, utopianism has to be tempered by pragmatism. As such, Greens are far better working with the grain of existing welfare systems than supporting the dismantling of those systems (Fitzpatrick with Caldwell, 2001). Secondly, social policy can remind environmentalists not to apply too generalised a critique to human destructiveness. At its worst, ecologism adopts an anti-humanist stance that treats all humans as the same. But although there may well be problems with a crude anthropocentrism, divisions between rich and poor (both nations and individuals) make it clear that environmental crises are more the product of a socio-economic system biased towards inequalities rather than the product of humanity *per se*. The trick is to recognise why environmental sustainability and distributive justice imply one another. Finally, many social policy commentators insist that the basic material needs of all must be provided for through universalistic provision. There is little point in preserving resources if the price is either scarcity for all or health and security for some. Sustainability must imply transforming the world in order to serve human well-being. If ecologists regard human and non-human well-being as irreconcilable then they may only be repeating the underlying mistake of productivist welfare.

These considerations also allow us to identify what Green social policies are *not*. For instance, they are not necessarily based upon the demand that public expenditure be reduced (George and Wilding, 1994: 170). This might be the demand of free market environmentalists but other Greens are likely to support existing or even higher levels of expenditure as a percentage of GDP. The demand is for growth measured in terms of GDP to evolve into other forms of measurement that either supplement GDP indicators or even replace them altogether (see Chapter 6). What this would do to the absolute level of social and public expenditure is difficult to anticipate but it certainly does not suggest that Green critiques be equated with radical Right ones.

The book

As indicated above, the objective of this book, and its companion volume, is to focus upon some of the themes that introductory treatments of environmentalism and social policy deal with too briefly or not at all. We are certainly pleased that ecologism has registered on the consciousness of many social policy academics but the latter risk underestimating the significance of the former if they imagine that it is just another 'welfare ideology'. Feminists have worked hard to challenge the assumption that gender issues could be simply bolted onto traditional welfare states and our hope is that these chapters will raise similar doubts *vis-à-vis* the environment. Therefore, although we cannot pretend that these publications are exhaustive we do hope that they open a debate at which previous treatments of the subject have only hinted.

Before offering a synopsis of the book there are several points to be made. First, 'ecologism' and 'environmentalism' are largely treated in this work as synonymous. We are aware that distinctions, with various degrees of subtlety, can be made between them but we have encouraged contributors to avoid jargon and long semantic diversions wherever possible. The ultimate rationale is that we have wanted many intellectual and political points of view to be represented, both in this book and in *Environmental Issues and Social Welfare*, but for contributors to avoid esoteric discussions of the meaning of concepts and terms. Secondly, contributors were asked for chapters that walked a line between the basic and the advanced. This means that for readers new to this subject the following chapters should provide accessible introductions but also analyses that reflect the authors' views and attempts to influence the relevant debates. Finally, we are aware that these chapters reflect a fairly narrow geographical perspective, one centred upon advanced welfare systems. There is obviously the scope to extend these debates towards those concerned with development studies, but this will have to wait for another time.

In Chapter 2 John Barry attempts to 'mainstream' the moral dimensions of Green thought in two senses. First, by arguing that many Green values and principles are already inherent within existing social practices and relations; second, by arguing that through avoiding the sirens of ideological purism we can recognise why the leap to a greener society may be less considerable than is often imagined. Indeed, there is an important distinction to be made between a sustainable society and a Green one. Whereas the former implies a restriction in the material demands we make upon a finite planet, the latter prescribes a particular form of morality that respects the intrinsic value of nature. Unfortunately, we usually miss this distinction due to the sharp and entirely artificial distinction between anthropocentrism and ecocentrism that continues to dominate the debate. Anthropocentrism rejects the idea of intrinsic value and so collapses the more radical versions of sustainability into the least radical ones, i.e. the discourse of sustainable development that politicians tend to favour; ecocentrism rejects the notion of instrumental value and so fails to compromise with more short term and practical forms of Green politics. If, instead, we were to adopt an 'enlightened anthropocentrism' then we could realise why Green politics implies a constant process of *critical vigilance* rather than the achievement of a once-and-for-all utopia.

A Green ethic therefore converges upon social policy because it is the latter which, for better or worse, dominates the politics of our unsustainable societies. Barry rehearses some of the arguments (see above) for developing an approach to social policy that is less concerned with economic growth, inequality and private consumption than the current welfare agenda. His main contribution, however, lies in a discussion of moral diversity, i.e. the ways of life and versions of the good that a greener society might embody. Must we sacrifice moral diversity to sustainability or are the two more compatible than both critics and supporters of environmentalism often imagine? For Barry, the question hinges on the notion of 'symbiosis', or the quality of the relationship between human and nonhuman. Strictly speaking, a sustainable society need not be symbiotic in that it could still imply the exploitation and misuse of the natural world. A Green society *would* embody symbiosis but, if interpreted in ecocentric terms, at the potential cost of destroying the gains of liberal democracy. Working through these problems requires us to re-think long-standing concepts such as 'welfare', 'development' and 'progress'. Barry's discussion of these issues is not intended to be the last word on the matter, precisely because the greening of state and society requires a theoretical open-endedness rather than the promotion of specific, ideological blueprints for social reorganisation. Therefore, although some might doubt whether it is as possible to distinguish between action and intention (and therefore between a sustainable society and a Green one) as Barry seems to believe, he undoubtedly establishes an argument that it is now impossible to avoid.

Mathew Humphrey discusses the meaning and significance of ideology at greater length in Chapter 3. Is ecologism an ideology in its own right or is it a set of ideas and aspirations that is too disparate to be labelled an ideology? Humphrey analyses Michael Freeden's view that ecologism is a 'thin' ideology at best, i.e. that it has only a limited core of ideas and conceptualisations that distinguish it from other ideologies. He agrees that ecologism's core is rather thin but finds that it is, nevertheless, more substantial and robust than Freeden's characterisation would suggest. This is because ecologism has new and important interpretations to propose regarding a series of 'adjacent' concepts that lie around the ideological core. For instance, it offers a distinctive perspective on human welfare, one where social well-being and environmental sustainability are interdependent. Since this interpretation of well-being is one that has not been developed by any non-Green ideologies it suggests that ecologism is more ideologically original than some would like to believe. Furthermore, this interpretation has practical policy implications. For if the environment requires greater protection than we have so far been willing to provide, this requires more of an emphasis upon satisfying the basic needs of all and less upon preference-satisfaction and material gain. Hence, the support which many Greens demonstrate for a Basic Income (see Chapter 8).

In Chapter 4 Tony Fitzpatrick turns our attention to the state and to democracy. How should we conceive the relationship between ecologism, democracy and social policy? Fitzpatrick proposes a theory of 'ecosocial welfare' that represents a conceptual space between Green social democracy and ecosocialism. By drawing upon an eclectic series of influences – liberal democracy, Marxism and post-structuralism – ecosocial welfare attempts to reconcile pragmatism with idealism, conservatism with radicalism. In one sense it agrees with Barry when he insists that Green politics must be multifaceted and open-ended, yet it also follows Humphrey in treating ecologism as representing an ideologically distinct point of articulation and negotiation. Ecosocial welfare theory therefore regards unambitious versions of Green liberal and social democracy as too self-limiting, yet neither does it make the mistake of demanding an all-or-nothing vision of a sustainable future.

The theory incorporates three principal elements. First, a 'radical reformism' that sets out to identify the contradictions and ecological inadequacies of our present social and welfare systems, and to encourage forms of bottom-up civic experimentation. Secondly, a 'discursive reflexivity' that draws upon ideas of performative politics and deliberative democracy. Finally, proposals for the collective control of natural and social resources that seek new forms of popular ownership disentangled from the statist abuses that have managed to discredit such objectives in the past. The theory of ecosocial welfare is explained here only in the most hesitant and tentative form, yet tries to offer conceptual resources to social policy that it currently lacks, a lack that is arguably inhibiting the subject from developing new approaches to the environmental challenges that we are already required to meet.

Michael Cahill address the importance of the local in Chapter 5, arguing that environmentalism can influence social policy to re-discover some of its pre-statist roots in an age when welfare services possessed more of a local character than they do at the moment. This is not to argue for an abandonment of the gains which centralisation and 'nationalisation' have brought, but it is to argue that if citizens are to be empowered then social policies will need to reinvigorate the local. Part of the problem, though, is that consumer capitalism, in its obsession with hyper-mobility, hollows out the local and abstracts us from the immediate and the proximate. Reversing this state of affairs, then, is not simply about devising new policies but reconfiguring our whole culture of consumerism and speed. It means re-valuing a moral consciousness of where we are for, if sustainable forms of welfare are to prevail, then our sense of ontological rootedness and emotional attachment must strengthen.

In Chapter 6 Tim Jackson explains why conventional measurements of well-being, those based upon economic growth indicators, are inadequate from an environmental perspective. The essential problem with indexes such as GDP is that they do not distinguish between creative and destructive forms of activity. If, in one example, I throw hand grenades into a shopping precinct then, so far as GDP is concerned, both the production of the grenades and the costs arising from the resulting carnage contribute to GDP growth. For Greens, the distinction between creative and destructive activities is particularly important and Jackson summarises the research which confirms their suspicion that the well-being of society has either not risen or, in some countries, actually declined over the last few decades. Over this period, environmental economists have laboured to develop alternative ways of measuring well-being that take into account a range of other factors, e.g. social inequalities and air pollution. Jackson reviews such attempts and acknowledges their superiority to GDP and traditional indicators. However, he also criticises these alternatives for still being too wedded to orthodox principles and assumptions; namely, those that focus upon 'material need-satisfiers'. If, by contrast, we were to recognise the validity of radical Green critiques then we could begin to move beyond our consumerist ways of life. The implication of Jackson's analysis appears to be that once we achieve a high and sustainable quality of life, that embodies the value of the non-material, then we will no longer be able to reduce well-being to a few statistics. Perhaps the more we experience welfare the less we are able to measure it.

Meg Huby explains the implications which the unsustainable use of resources has for social policy in Chapter 7. That those on low incomes suffer from energy and water poverty has long been noted by social policy academics given the fact that energy and water are two of the basic needs that human life requires. Traditionalists have usually found the solution to be in either reducing the cost of these goods and/or of raising the incomes of the poorest. Yet according to Huby we cannot make effective recommendations until we have debated the ecological consequences of such solutions and decided how the social and environmental dimensions can be made to work together. The dilemma is how to ensure that the short term alleviation of poverty does not exacerbate the long-term ecological problems that we face. Indeed, only if we accept that poverty-reduction has an environmental component are we likely to adopt policies that facilitate the convergence of social and environmental objectives in the long term. Without such convergence, poverty reduction in the near future may be short-lived and even counter-productive.

Huby applies her expert knowledge of the subject to an analysis of existing policy developments and recommends how those policies can be improved upon. Her basic conclusion, though, is that the trend towards deregulation and privatisation that has characterised government policy over recent years must be replaced by a regulatory regime which is oriented to the maintenance of public goods and where a sense of public ownership is restored.

This conclusion is one that also inspires the final four chapters of the book, each of which applies the lessons of environmentalism to some of the key areas of social policy. Chapter 8 deals with what is often taken to be the archetypal Green social policy proposal: Basic Income. Fitzpatrick offers an overview of the relevant debates and summarises the reasons why Greens might want to support a Basic Income, as well as the reasons why they may not. He shows that *if* ecologists have reason to support the Basic Income proposal then they can only do so if Basic Income sits comfortably and effectively within a wider policy reform package. Indeed, many Greens have recognised this themselves and it is rare to find Basic Income discussed without some discussion of informal economic exchange, working-time reductions and eco-taxation. These are the subjects of, respectively, Chapters 9, 10 and 11. Therefore, just as Basic Income may represent the axis of a Green approach to social welfare, so Chapter 8 represents a starting-point for the book's remaining chapters.

The first of these subjects has been an enduring theme within social policy and has gone under a variety of names, e.g. social economy, third sector, civic association, cooperation circles. What such terms refer to is a space within civil society that is not reducible to the relations of state, market or family/kinship. Each of these interacts with this space in a complex series of relations but none quite captures what is unique and significant about it. Social policy commentators have usually fallen back upon a nomenclature such as the independent or voluntary sector; yet this, too, does not quite capture the depth, complexity, spontaneity and ordinariness of this space.

The social economy, for want of a better alternative, is the subject of Chapter 9. Chapter 8 observes that Greens look beyond the formal to the informal economy of value and exchange, and Colin Williams analyses that informal economy by summarising recent research into Local Exchange and Trading Systems (LETS). LETS have been fashionable among policy pundits over recent years. For what Williams calls mainstream commentators, they provide the opportunity to reconnect the excluded with the world of work; for Greens, they represent an alternative to the mainstream economy, a point of leverage through which broader notions of social participation could be encouraged. Williams concludes that the implications of LETS are broad enough to satisfy both types of commentator, but that without an emphasis upon the 'social economy' aspects of LETS then such initiatives are likely to be under-utilised. Unfortunately, mainstream policy continues to emphasise the narrower definitions of inclusion and exclusion, ones where paid employment is the norm. If, therefore, ecologists can encourage a widening of our conceptual horizons then we may be able to reform welfare systems around the social economy in a much bolder, less conservative manner than ever before.

The subject of working time reductions is picked up by Adrian Little in Chapter 10. Little traces the fashion for a politics of free time in the traditions of left-libertarian and ecological thought, despite the fact that many Greens and socialists have supported paid employment as a vehicle for social and moral transformation. This politics has increased in influence with the advent of a 'new capitalism', where a globalised economics of flexibility can no longer shore up the ideology of paid work. Little underlines the importance of Andre Gorz in this respect, since Gorz has long maintained that reductions in working hours do not have to be paid for with concomitant reductions in wage levels. If we had the political will to channel increases in productivity in more socially valuable directions then we could have more free time whilst maintaining present (material) standards of living. Elements of that political will have emerged in recent years in France where the Jospin government made a concerted effort to reduce working time without the economic earthquakes that orthodox critics have predicted. Little acknowledges that the French example is neither particularly Green nor libertarian, but recommends it as an example that could lead towards more radical strategies. And although he draws attention to the dilemmas that Greens must overcome before such potential can be exploited, Little is clear that Greens must and should have a considerable contribution to make to the economics of the future.

James Robertson closes the book by providing a succinct but effective overview of eco-taxation in Chapter 11. According to Robertson, conventional sources of public revenue are no longer tenable due to globalisation and technological advances, nor are they particularly desirable since they tax individuals for the value that they add to society rather than for the value which they subtract. Therefore, public expenditure in the future should be based far more upon the taxation of common resources such as land and energy. Robertson uses just one example of a common resource, the issuing of new money, to illustrate how much could be raised by this measure but is currently lost due to the illogical nature of present arrangements. He also addresses the issue of whether eco-taxes must be regressive and insists that, although they certainly contain regressive features, a Green political economy is capable of more than compensating for these features and so of retaining the traditional radical demand for progressive policies and social justice. Robertson concludes by applying these lessons to the global dimension, arguing that a new political economy, and a Green welfare state, can only be based upon a re-conceptualisation of distribution and of market forces.

Robertson's analysis is self-consciously utopian, while being rooted in the work of a lifetime that has been dedicated to finding new approaches to economic and social policy. As he himself observes, the line between utopian thinking and practical action, between what 'pragmatists' reject today and what they claim credit for tomorrow, is so blurred that radicalism is often victorious without having to be triumphalist. Without the utopian imagination, pragmatism would lose its rationale, condemned to rowing in ever decreasing circles; with it, pragmatism can afford to be more ambitious, ranging across an ever widening circumference whilst keeping sight of the present. This observation returns us to the point John Barry makes in Chapter 2 about the necessity of intellectual pluralism in Green thinking. Both Barry and Robertson call for an idealistic realism, for a rigorous but nevertheless wide-ranging discourse out of which Green social policies can begin to form. If this call is heeded then, indeed, the greening of the welfare state becomes not only a project for the future but a project that develops in the here and now, in the midst of where we already are. If you can remember to bear this in mind then it is time for you to read on.

2
The Ethical Foundations of a Sustainable Society

John Barry

Introduction

How are we to live? What sort of society do we wish to live in? What is the 'good life' for humans? What is the 'just society' and how can we create it? These are some of the long-standing ethical questions we have asked ourselves throughout history in all human societies. Every normative political theory and set of policy prescriptions provides answers, whether explicitly or otherwise, to such questions, and so also provides an ethical theory or set of guidelines about how individuals ought to live and how society ought to be organised. Green political theory is no exception and, as outlined below, its particular view on these ethical questions makes it unique among contemporary political theories. First, Green political theory offers unique answers to many of the longstanding ethical issues indicated above. Secondly, Green political theory is concerned with new areas of moral concern (particularly in regard to how we ought to treat the nonhuman world), areas about which there has, until recently, been little or no debate. Thirdly, and related to the previous two points, Green political theory thus has a distinctive approach to, and perspective upon, social policy and the welfare state, and so has a distinctive position in any discussion of the 'greening' of these.

In exploring the ethical foundations of a Green society I will concentrate on the contribution that Green moral and political theory can make to the ethical principles and ideas underpinning sustainability. However, I do not presume that Green normative theory has a monopoly on the debate about the ethical basis of a sustainable society. Nor do I believe that such ethical foundations necessarily have to conform to those propounded by the Green moral and political position. While

I do think that the ethical basis of a Green society will be related to, or derived from, many of the claims and positions made by Green theorists, parties and activists, this chapter also presents some arguments which are either derived from non-Green perspectives or at odds with some of the prescriptions often advanced by Greens. Indeed, for this reason I will talk of the ethical foundations of a *sustainable* society rather than a *Green* one, so that the discussion is not confined to a narrow ideological arena. *While a Green society will necessarily be a sustainable one, it is not the case that a sustainable society necessarily has to be a Green one.* The aim of this chapter is thus to explore the ethical foundations of a sustainable society and the part Green moral and political principles will or can play in terms of them.

Thus, while the ethical basis of a more sustainable society will be different from that which operates in modern day liberal democratic societies, it is not the case that this basis will be completely alien, unintelligible and unrelated to the common ethical commitments of people currently living in these societies. For example, while I do think that the ethical basis of a Green society would require a less intensive meat-eating food economy and culture than that visible in the developed world, this need not be derived from any putative Green moral principle about compulsory vegetarianism on the basis of animal rights. There are other ethical principles and perspectives we can call upon to demonstrate the immoral and ethically suspect practices of the modern western, intensive meat industry.

There are two broad reasons why I think the ethical foundations of a sustainable society will have much in common with many of the shared moralities of modern societies.

The first concerns ethics, in that while current societies are unsustainable and permit many morally indefensible ab/uses of the nonhuman world (most noticeable in the treatment of animals), it is not the case that individuals are wantonly or *wilfully* abusing the nonhuman world. Often, the ethical motivations behind current unsustainable and unjustified uses of the nonhuman world are based on ignorance rather than maliciousness. Indeed one of the dominant ethical impulses behind many of the ways in which society uses/abuses the nonhuman world is a desire to improve the human condition. One of the biggest problems in persuading people of Green moral and political arguments is that many find it hard to accept that actions motivated by good intentions, e.g. the desire for greater material comfort, can lead to morally bad consequences. This is even harder when it is the case that morally bad consequences are a *necessary and inevitable* rather than a *contingent* outcome of such behaviour. As a result, it is only recently that the Green critique about unsustainable economic growth, and its deleterious effects upon both human welfare and the well-being of the nonhuman world, have begun to infiltrate popular and political consciousness. If the immorality lies within the ignorance and not within the intentions then there is no *a priori* reason for Greens to reject the latter.

The second reason concerns some practical issues. If the ethical foundations of a sustainable society are associated too closely with an ideologically 'Green' or ecocentric perspective then the chances of a democratic transformation to a more sustainable society are severely compromised (J. Barry, 1999a: ch. 2). Employing Rawls' distinction between 'political' and 'metaphysical' bases of political morality, ideologically Green and deep ecological views cannot underwrite a political agreement around sustainability. According to Rawls (1985: 225), 'as a practical political matter no general moral conception can provide a publicly recognized basis for a conception of justice in a modern democratic state'. Thus, policies for sustainability must be expressed in terms of support from a broad spectrum of existing substantive moral/ethical approaches and views. As Goodin (1992: 18) suggests, if a sustainable society demands a completely alien set of 'Green' values, then the Green project is even more hopelessly utopian than its worst critics suggest. In short, if we wish to sustain the moral pluralism and multicultural character of modern society, as well as sustaining the environment, then it simply cannot be the case that a sustainable society must be based on the widespread belief and acceptance of specifically 'Green' morality and way of life.

My starting point is that the ethical foundations of a sustainable society will not and should not be based on any one master ideology or set of moral prescriptions. Rather, my view is that they will necessarily include 'non-Green' elements and, indeed, include ethical commitments that may go against aspects of many Green aims and objectives. Thus, I suggest that the ethical foundations of a sustainable society will and ought to be pluralistic, drawing upon different ethical traditions and frameworks. My suggestion is that sustainability does not require an exclusively 'Green' morality and ideology, and insisting on the latter would undermine the moral pluralism which is an essential part of the freedom and diversity of modern societies that a sustainable society would wish to preserve. In other words, if we tie the achievement of a sustainable society too closely to a 'Green' moral worldview, then we may end up sustaining a social order which achieves sustainability by sacrificing widely valued aspects of liberal democracy. Moral diversity should not be sacrificed to biological diversity.

Thus, one does not have to be a 'Green' or indeed accept ecological or Green moral or political principles in order for one's actions to have sustainable outcomes. The values and belief systems of social actors are less important than the consequences which their actions have for sustainability. That the same sustainable outcome or policy can be supported by, or be consistent with, a variety of different ethical worldviews and value systems is something that makes a sustainable society a realistic possibility within modern, multicultural and morally diverse societies.

Green ethical principles

Many of the distinctive and radical ethical principles of Green theory focus on the relationship between humans and the nonhuman world, those 'ecological' dimensions of human life, society and action. Some examples of these ecological moral issues include:

- Does nature have intrinsic value or is its value solely related to its instrumental value to humans?
- Is meat-eating justified and, if so, under what conditions?
- Do animals or other parts of nature have rights?
- How should humans value ecosystems?
- What obligations do we have to non-sentient life?
- How do we regulate and decide conflicts between human and nonhuman interests, welfare or well-being?

This last question will here be taken as the most salient and overarching issue that needs to be discussed and developed in sketching out the ethical foundations of a sustainable society.

Green political theory has challenged us to create new discourses and 'communities of justice' based on new morally significant others, the interests of which have to be taken into account when making decisions that affect them. Thus, the ethical foundations of a sustainable society can be summarised as relating to the following four varieties or dimensions of justice:

- social justice, i.e. justice among the current generation of humans;

- interspecies justice, i.e. justice between humans and the nonhuman world – its living beings, systems, non-living entities and processes;[1]
- global justice, i.e. justice within the current generation of humans across different societies;
- intergenerational justice, i.e. justice between current and future human generations.

While the dominant discourse of social justice is familiar, it is in extending our moral concerns in time (to future generations), in space (to non-citizens) and, of course, beyond the species barrier (to nonhumans), that we can see the outline of the distinctively Green elements of the ethical basis of a sustainable society. This is not to say that all actions and behaviours in a sustainable society must be motivated by such concerns, any more than behaviour in a liberal society must always be motivated by liberal values. However, before moving on to this point it is necessary to briefly discuss one approach which exemplifies how *not* to derive the ethical basis of a Green society.

Reading off ethical principles from nature

Some versions of Green moral theory, most noticeably ecocentric or deep ecological theories, are often exclusively or largely concerned with developing moral principles to govern relations between the human and nonhuman worlds, and have little to say about the sort of moral principles which ought to govern human social relations (J. Barry, 1999a). They thus provide only a partial answer to the questions we are examining.

More worrying is the tendency often found in these versions of Green moral theory to 'read off' ethical and political principles for human behaviour and society from the natural world (J. Barry, 1994, 1999b: ch. 1; de-Shalit, 2000: 99–103). This 'reading off' method claims that we can better describe or explain human society by applying the knowledge derived from the study of the nonhuman world, since we too are part of nature, i.e. a particular species of animal. But alongside this 'descriptive' claim, it also purports that the study of the nonhuman world can have prescriptive power. In short, the reading off hypothesis states that we can both describe the social world as it is and prescribe how it ought to be from the application of knowledge gained from the study of the natural world. We can 'read off' how human society is and ought to be from looking at the nonhuman world, and applying the governing principles we find in the non-human world and translating them into ethical principles governing human society. Thus Dobson (1990: 24) writes that for 'ecologism' there is a

> strong sense in which the natural world is taken as a model for the human world ... . the principal features of the natural world and the political and social conclusions or prescriptions that can be drawn from them are:
>
> diversity   – toleration, stability and democracy;
> interdependence   – equality;
> longevity   – tradition;
> nature as 'female' – a particular conception of feminism.

However, reading off how human society or social relations ought to be from an examination of the natural world inevitably involves the projection of social claims, aims and positions onto that natural world. That is, we 'read off' from the natural world what we project or 'read into' it (J. Barry, 1999b: ch. 2). For some, the natural world is a place of harmony, co-operation and balance which allows them to posit and justify socially equivalent ethical principles, such as democracy, social justice and equality. However, for others 'nature is red in tooth and claw', a place of competition and 'survival of the fittest', allowing them to posit a set of ethical principles which completely contradicts the previous set. Which reading is correct? How are we to judge which reading and related set of ethical principles is the correct one to use as the ethical foundation of a sustainable society?

The short answer is that there *are* no determinate readings of the environment because there are no value-free readings and the whole idea of reading off ethical principles from nature is deeply flawed. We need to ask why *any* reading of the natural environment should be seen as somehow authoritative in prescribing ethical or other principles for governing human society. Nature may be harmonious or competitive or both but, we can ask, so what? Why should we necessarily seek to derive principles for governing human society from the natural world? The question we should ask is not what particular ethical principles we can derive from or discern in nature, but rather the more fundamental question of why seek to do this in the first place? One of the dominant and problematic ways in which radical Green theory buys into this 'reading off' hypothesis is in insisting that only *ecocentric* moral views and motivations can constitute the ethical foundations of a sustainable society. It is to an examination of the relative merits of ecocentrism and anthropocentrism that we turn to next, therefore.

Ecocentrism vs. anthropocentrism

A central aspect of Green political theory is a concern with the normative status of the nonhuman world and its treatment by humans.
Within the literature there are two extreme poles that, together, create
a continuum along which different Green normative positions can be
plotted. At one end of this continuum there is what can be termed an
'arrogant anthropocentrism' (Ehrenfeld, 1978). This normative
position holds that the world is essentially meaningless and its only
value consists in the instrumental value human beings project onto it
when fulfilling their ends. Its philosophical origins lie in a variety of
sources, chief amongst which is the theological idea that the world has
been made for our use and enjoyment (White, 1967). Eckersley (1992:
51) defines it as 'the belief that there is a clear and morally relevant
dividing line between humankind and the rest of nature, that
humankind is the only or principal source of value and meaning in the
world, and that nonhuman nature is there for no other purpose but to
serve humankind'. This arrogant anthropocentrism is, from the Green
perspective, part and parcel of the modern world view.

At the other end of the continuum we find what may be termed
'extreme ecocentrism', most commonly associated with deep ecology.
Part of the rationale to find a non-anthropocentric moral framework was
based on the idea that part of the cause of the ecological crisis was to be
found in the anti-environmental attitude expressed by this arrogant
anthropocentrism. As such, the solution to the crisis was assumed to be
found in its opposite, namely a non-anthropocentric or ecocentric moral
perspective. The first modern expression of this ecocentric position can be
found in the seminal paper by Arne Naess. Naess (1973) outlined one of
the most well-known and developed strands of ecocentric thought,
namely 'deep ecology'. Whereas both arrogant anthropocentrism and
what Naess called 'shallow ecology' (whose primary concern was the
effect of ecological damage on human health and well-being) viewed
nature as possessing instrumental value, deep ecology held that nature
should be seen as having intrinsic value and should be protected for its
own sake and not simply because it was of benefit to human beings.
While there have been many developments within the deep ecology
position since Naess's original formulation (see J. Barry, 1994), deep
ecology still holds to the basic ecocentric position that anthropocentrism
is part of the ecological problem and needs to be replaced with a more
ecocentric or earth-centred moral perspective.

The deep ecology position is extremely radical for it demands not
simply the protection of nature but also that this protection be done

for the right (ecocentric) reasons. For example, from a fully deep ecological point of view, not only must we preserve the Amazonian rainforests, but we must do so because the rainforests have intrinsic value and ought to be protected for their own sake and not for any human benefit. Thus an argument for the protection of the rainforests on the grounds that it would be prudent to do so (because the rainforests may contain important medical substances) would fall short of the deep ecology standard since the reasons for protecting nature would be ones based on human interests. For deep ecologists, the problem with such anthropocentric moral reasoning is that arguments for the protection of nature premised on human interests mean that nature does not enjoy secure protection, since human interests may easily change from protection towards development or exploitation.

However, the difficulties associated with the deep ecology position, not least of which is how to translate its philosophical arguments into practical policies, and the seemingly endless disputes within Green political theory between 'deep' and 'shallow' perspectives (J. Barry, 1994), have led to an attempt to combine the deep ecology critique of arrogant anthropocentrism with a more defensible Green moral position than that of ecocentrism. In essence, the starting point for this new Green moral framework is to argue that the mark of Green moral theory is not that it be some form of ecocentrism but rather that it displays a critical attitude towards anthropocentrism (Hayward, 1997; J. Barry, 1999a; Baxter, 1999; de-Shalit, 2000; Radcliffe, 2000). This recent development within Green normative theory agrees with the ecocentric aim of criticising the excesses and negative effects of arrogant anthropocentrism, but disagrees with the ecocentric aim of replacing anthropocentrism altogether. Rather, the real target of the ecocentric critique should be the *arrogance* not the *anthropocentrism*. In other words, there may be forms of anthropocentrism which are compatible with Green goals, forms that may be indispensable if Green aims and values are to be realised.

One of the main problems with the ecocentric position is that it demands a complete change in the worldview and ways of thinking of those populations towards which the Green message is addressed. While it may be desirable for human culture and consciousness to become more ecocentric over the long term, it may also be that such a complete change in the way people think and act is unnecessary or even an obstacle to achieving Green goals in the short term. So, criticising the *arrogance* of dominant anthropocentrism may imply encouraging a symbiotic relationship between human society and nature that deep ecology, with its over-valorisation of the latter, cannot conceptualise. It is to a discussion of this symbiosis that we now turn.

Between sustainability and symbiosis

I will use the terms 'sustainability' and 'symbiosis' to describe the main dimensions of the ethical basis of a sustainable society. Green politics suggests that those ethical foundations would seek to encourage what may be called 'symbiotic' rather than 'parasitic' human-nonhuman relations (J. Barry, 1999a: ch. 2). 'Symbiosis' and 'parasitism' convey the general moral nature of social-environmental relations, while sustainability and unsustainability convey the material exchanges between society and environment (but which also convey moral relations between humans). There is no assumed 'mapping' of these two sets of criteria such that morally symbiotic social–environmental relations are also sustainable, and vice versa. It is perfectly possible for a society to be morally parasitic *and* ecologically sustainable. Thus while a society may be sustainable (in the sense of not violating long term material relations between society and environment), it does not necessarily have to be 'Green' in this ethical sense of symbiotic relations between human and nonhuman world. A society is describable as Green if, in addition to achieving sustainability, it also has elements of symbiosis in its treatment and use of the natural world.

This gives us the following sets of relations between human society and the natural world (see Figure 2.1). As the term suggests, parasitic

Figure 2.1   Human society and the natural world

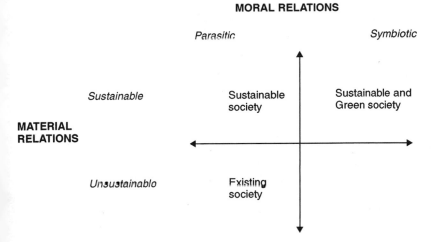

MORAL RELATIONS

Parasitic   *Symbiotic*

| *Sustainable* | Sustainable society | Sustainable and Green society |

MATERIAL RELATIONS

| *Unsustainable* | Existing society |

moral relations between human society and the natural world denotes a relationship of use in which humans exploit and abuse the natural world. Humans in a parasitic relation with the natural world simply view it and its entities as a store of resources or means to fulfil human ends, decisions on how to use these resources are not based on any sense that relations between humans and nature ought to be governed by anything more than technical or prudential considerations.

It is important to point out that the 'symbiotic ethic' proposed here, as the overarching moral basis for governing human-nonhuman relations in a sustainable society, is not an ecocentric ethic as that term is conventionally understood (see above). The symbiotic ethic is an anthropocentric 'ethic for the use of the environment', as opposed to an 'ecocentric ethic' (Regan, 1982) where this is premised on non-anthropocentric reasons for the protection of nature. In this way, one could say that the ecocentric ethic is more ideologically suited as the ethical foundations of a 'Green' society, whereas an ethic of use is most appropriate as a part of the ethical foundations of a 'sustainable' society.

The central claim of an ethics of use is that an extension of human interests can achieve many of the practical outcomes desired by non-anthropocentrists but on a more secure basis. This position starts from Norton's (1987: 222) observation that, 'A narrow view of human values ... encourages environmentalists to look to nonhuman sources of value to justify their preservationist policies.' A broader view of human values and interests in the world may obviate the necessity for non-anthro-pocentric sources of moral concern. Part of this broadening process involves the examination and extension of human interests in the environment, including environmental interests based on a concern for future generations. The reason for this is that this critical and self-reflexive process opens up the possibility of new moral relations between humans and nature within anthropocentric moral reasoning. The problem with most critiques of anthropocentrism is that they are insufficiently sensitive to its environmental possibilities, especially in relation to the political defence and articulation of Green policies and ideas. In this way, I suggest that the ethical foundations of a sustainable society are not ecocentric, but will be focused on the creation, encouragement and maintenance of symbiotic uses of the nonhuman world.

Although an ethics of use is by definition human-centred and related to human interests, it also argues that the justification of a particular environmental practice merely on the grounds that it fulfils a human interest is no longer acceptable. The fact that a particular use of nature fulfils a human interest cannot be taken as a decisive reason either for its initiation, continuation, or its continuation in the same manner. It is *this* understanding of anthropocentrism that deserves to be criticised as 'arrogant humanism', the idea that the mere reference to human interests is sufficient to morally justify any human use/abuse of nature. An ecological ethic of use argues, on the other hand, that human interests are a necessary *but not a sufficient condition* for the justification of human–nature relations. For any human–nature relation to be fully morally justified the particular interests which that relation fulfils must be justified. This position begins from Midgley's (1983: 17) conviction that, 'however far down the queue animals may be placed, it is still possible in principle for their urgent needs to take precedence over people's trivial ones'. In short, not all human interests, simply by virtue of being human, are equally acceptable or justifiable. That we must consume parts of nature to flourish, and use it in other ways to mark human life, does not mean that all uses are equally justified. Some are more morally defensible than others. The ethical basis of a sustainable society then becomes centred on determining defensible or permissible human uses of the nonhuman environment, and distinguishing these from impermissible, trivial and unjustified abuses.

Weak or enlightened anthropocentrism

If we accept that the dominant ethical worldview within modern societies is broadly anthropocentric, and we also accept that the ecocentric critique is misplaced in that it conflates a particular conception of anthropocentrism (namely an arrogant one) with anthropocentrism *per se*, then we can say that what marks the ethical basis of a sustainable society is a critical attitude to and rethinking of anthropocentrism rather than its rejection and replacement with an ecocentric or deep ecological worldview (J. Barry, 1999a). What I suggest is that a reformed and enlightened anthropocentric ethic offers a more defensible and attractive moral basis than ecocentrism. Adapting to my own purposes a distinction Norton (1984) draws between 'strong anthropocentrism' and 'weak anthropocentrism' I now outline the way in which an ecological ethic of use turns upon the distinction between human preferences and interests as well as the extension of human interests.

According to Norton (1984: 134), 'A value theory is strongly anthropocentric if all value countenanced by it is explained by reference to satisfactions of felt preferences of human individuals'. For my purposes the distinguishing feature of strong anthropocentrism is the claim that human-nature relations can be justified by reference to felt or 'given' preferences alone. From the strong anthropocentric position it does not make sense to talk about moral judgement derived independently from human preference-fulfilment. It is the reductive character of strong anthropocentrism in conceiving of human–nature relations in purely instrumental terms (typically economic ones) that 'crowds out' both the need for their justification and the requirement that moral considerations ought to act as side-constraints on how relations are managed. Strong anthropocentrism, in holding that the moral justification of human uses of the natural world need not go beyond reference to preference-fulfilment, reduces the various human interests that extend (or could extend) over our relations with nature to the preferences humans currently happen to have. According to Norton, if preferences are insulated from critical appraisal, human interests in the nonhuman world are in danger of becoming narrower than they might otherwise be.

At root, this critique shares with critical theory the view that economic reasoning, if unchecked, can lead to the 'demoralisation' of human–nature exchanges (as well as a demoralisation of human relations). Green politics is thus not against economic reasoning but does seek to place it within its proper context as one among other modes of human interaction and valuation. From a Green point of view, the aim must be to assess preferences by reference to the 'seriousness' or 'worthiness' of the human interest to which it is related. An ethics of use is thus a form of 'weak anthropocentrism', which acknowledges a plurality of human interests in the natural world and a variety of permissible, justified and morally appropriate relations to that world. It explicitly sets itself against any one 'true' or dominant ethical view of our relation to and use of the nonhuman world, which both strong anthropocentrism and strong ecocentrism tend to do. This ethics of use acknowledges the 'tragic' dimension of human 'being in the world' in the sense of fully recognizing that, as in other areas of human life, human–nature affairs are often characterized by a clash of competing and mutually exclusive 'goods'.

An ethics of use is concerned with establishing the contested boundary between legitimate use and abuse, as well as the often more complex issue of when use cannot be morally justified at all, i.e. the line between use and non-use.[2] These boundaries can never be fixed in the manner that an *a priori* commitment to the preservation of nature would require since they are inevitably contingent. The relative indeterminacy of this ethics of use will not satisfy those who seek cast iron protection of the nonhuman world. However, there is simply no remedy for this, and the most we can do is to acknowledge and articulate that fact. This is the moral responsibility at the heart of the ethical foundations of a sustainable society, in that a key part of the latter is 'monitoring' this indeterminacy and identifying, from case to case, when and why human use of the environment becomes unjustified abuse. For example, the legal protection of animals from particular forms of abuse and cruelty we find in many countries serves as an example of this type of ethical monitoring, and which also serves as a real-world example of how the (welfare) state can be seen to protect the interests and welfare of the nonhuman world.

Sustainability and intergenerational justice

We do not inherit the earth from our parents but borrow it from our children.

Another strand of a non-ecological ethical foundation can be found in those who point out that a concern for future generations may result in environmental policies whose outcomes would result in sustainable human–nonhuman relations, but without being based on any direct concern for nonhumans.

The following quote from Jacobs neatly outlines both the general contours of the concept of sustainability and how it directly relates to relations between current and future generations. According to Jacobs (1996: 17):

The concept of 'sustainability' is at root a simple one. It rests on the acknowledgement, long familiar in economic life, that maintaining income over time requires that the capital stock is not run down. The natural environment performs the function of capital stock for the human economy, providing essential resources and services. ... Economic activity is presently running down this stock. While in the short term this can generate economic wealth, in the longer term (like selling off the family silver) it reduces the capacity of the environment to provide these resources and services at all. Sustainability is thus the goal of 'living within our environmental means'. *Put another way, it implies that we should not pass the costs of present activities onto future generations.* (emphasis added)

Those who seek to criticise the arrogance of the dominant form of anthropocentrism and seek to base the Green position on an enlightened alternative may be said to agree with Norton's (1991) 'convergence hypothesis'. For Norton, in terms of environmental policy outcomes, there is a convergence between the ecocentric position, as represented by deep ecology, and a far-sighted anthropocentrism which is concerned about our obligations to future generations (cf. Fitzpatrick, 2001b). According to Norton (1991: 227):

> introducing the idea that other species have intrinsic value ... provides no operationally recognizable constraints on human behaviour that are not already implicit in the generalized, cross-temporal obligations to protect a healthy, complex, and autonomously functioning systems for the benefit of future generations of humans. Deep ecologists, who cluster around the principle that nature has independent value, should therefore not differ from long-sighted anthropocentrists in their policy goals for the protection of biological diversity.

This convergence between a concern for future generations and a concern for the nonhuman environment is the essence of the debate around sustainable development and ecological sustainability. Thus the recent debate within Green moral theory is largely between those who seek to develop a 'weak' or 'reflexive' anthropocentrism as the normative basis for Green politics (J. Barry, 1999a; de-Shalit, 2000; Hayward, 1997) and those who still hold to the belief that only an ecocentric moral basis can both sustain a Green politics worthy of the name and stand as the ethical basis of a sustainable society (McLaughlin, 1994; Naess, 1995).

Social justice and economic growth

In terms the of ethical principles covering human social relations, egalitarianism and social justice figure strongly in the ethical foundations of a sustainable society. A commitment to lessening socioeconomic inequalities is crucial for a number of issues. First, if a more sustainable society necessarily requires less economic growth, as many Greens suggest, this undermines one of the central justifications for inequality, i.e. the standard argument that inequalities of income and wealth are necessary in order to promote economic growth. Second, as the growing theoretical and empirical research on environmental justice demonstrates, poor people tend to live in poor and degraded environments, and improving their environmental and socioeconomic standards requires reductions in social inequality.

In the welfare states of the west, state legitimacy is in part dependent upon a commitment to reduce inequalities via redistributive measures. However, co-existing with this is the above argument which claims that economic inequality is necessary for creating the conditions of economic growth by encouraging individuals to work harder and be more productive, thus increasing the overall amount of wealth in society. In other words, an unequal distribution of the benefits of socially produced wealth is a necessary condition of a growing, successful economy (less is said about the unequal distribution of the costs and risks of economic performance). Wealth and income inequalities are argued to be necessary economic incentives for economic productivity and growth. The state can then tax this growing wealth to fund its welfare activities and so make that economic growth socially just. The basic argument is that while some gain more than others, everybody gains, or as Adam Smith put it over 200 years ago, 'A rising tide raises all boats'.[3]

The argument for a more sustainable society can thus be linked to a critique of the social inequalities that are a structural component of contemporary social and economic policies aiming to produce maximum economic productivity and growth. With an economy not geared towards maximising production, income and paid employment, the justification of an unequal distribution of socially produced wealth cannot be that it is required for procuring greater wealth production. In short, with the shift to a less growth-orientated society, the normative basis for social co-operation needs to be re-negotiated, as does social policy (J. Barry, 1998). The welfare implications of this are dramatic given that one of the central justifications for social policy is to lessen socioeconomic inequalities via the redistribution of income, goods and services generated from a growing economy. The argument is that if one wishes to reduce inequalities then accepting ecological limits to growth may be a more realistic way of achieving it, since demands for a more egalitarian distribution of social wealth are more likely to be made in a non-growing economy.[4]

A less radical Green position on economic growth is taken by Jacobs (1996) who makes the important distinction between economic growth understood as increases in personal disposable income (which is the dominant view) and economic growth as implying greater public spending and investment. As he puts it:

A more sustainable economy would have higher investment; *it is unlikely to have higher private consumption.* Where additional

> consumption is required it will often have to be in the public sector, on goods such as public transport, environmental protection, health care and education ... *sustainability probably does mean that the era of taken-for-granted exponential consumption growth is at an end.* (Jacobs, 1996: 33–5; emphasis added)

The point is that economic growth orientated towards more public forms of investment and consumption does seem to imply less overall consumption (and associated production-related ecological damage) than economic growth aimed at increasing private consumption, income and wealth. Examples of this would be to achieve a balance between current individualised consumption patterns with more collective forms of production service provision, e.g. public transport. The aim would not be to transcend or abolish individualised consumption but to balance this out with public, collective (and not necessarily state-centred) modes of production and consumption. In this way, a shift towards more public patterns of production, consumption and investment would help to create a more sustainable economy as well as making a contribution to reducing socioeconomic inequalities within society.

Thus one could imagine a 'sustainable' social policy in which the aim was to provide goods and services which were either produced and distributed in an ecologically sustainable manner and/or produced and consumed by a variety of units ranging from the individual to households and collectives. The aim here would be to move from the co-existence of '(unequal) private wealth and public squalor', which many feel characterises modern, unsustainable industrial societies, towards a less unequal distribution of private wealth as part of a shift towards more public-orientated forms of economic growth and associated consumption, production and investment patterns.[5] In this way, the ethical basis of a sustainable society include a firm commitment to social justice, the public sector and greater equality.

Moral diversity

Many believe that a sustainable society would be characterised by less variety than that pertaining in existing societies.[6] In short, for many critics of sustainability, the preservation of biodiversity will be at the cost of less diversity in social ways of life. This is an anti-sustainability argument sometimes used alongside the more common critique that a sustainable society would result in a lowering of human welfare and standards of living. Here, the arguments that a sustainable society will be less resource and energy intensive is taken to mean that the resources available to individuals to realise their diverse life plans and views of the good will be considerably less, thus reducing the plurality and diversity of lifestyles in a sustainable society. In addition, there is the suspicion that a sustainable society necessarily implies the state promoting some ways of life and proscribing others, thus violating its impartiality and neutrality *vis-à-vis* views of the good life. So a sustainable society is charged with transgressing the liberal principle of state neutrality in that it would require the state to interfere directly or indirectly in the life plans and chosen modes of life of individuals and groups.

While I think it the case that particular forms of life based on intensive inputs of energy and resources and extensive outputs of pollution will most certainly not be consistent with a sustainable society, this does not mean that the overall amount of diversity in human modes of living need necessarily decrease. In 'Of the Stationary State', Book IV, Chapter VI of his *Principles of Political Economy*, first published in 1848, Mill suggested a radical critique of received wisdom concerning the progress of society. He suggested that the desire for more and more material goods and services, based on the domination of nature and the more intensive use and application of science and technology, was perhaps a too narrow view of 'social progress'. What Mill (1900: 453) had to say is worth quoting in length:

> I cannot, therefore, regard the stationary state of capital and wealth with the unaffected aversion to it so generally manifested towards it by political economists of the old school. I am inclined to believe that it would be, on the whole, a very considerable improvement on our present condition.... . It may be a necessary stage in the progress of civilization. ... But the best state for human nature is that in which, while no one is poor, no one desires to be richer, nor has any reason to fear being thrust back, by the efforts of others to push themselves forward. ... It is only in the backward countries of the world that increased production is still an important object: in those most advanced, what is economically needed is better distribution, of which one indispensable means is a stricter restraint on population.

In Mill's eloquent defence of the 'stationary state', in which wealth, capital and population are held as constant as possible in a non-growing economy, we find one of the earliest outlines of a greener society. To those who suggest that a non-growing economy would be a social disaster, Mill (1900: 455) replies that

> It is scarcely necessary to remark that a stationary condition of capital and population implies no stationary state of human improvement. There would be as much scope as ever for all kinds of mental cultures, and moral and social progress; as much room for improving the Art of Living, and much more likelihood of its being improved when minds are ceased to be engrossed by the art of getting on.

Thus it is an empirical question whether overall welfare would decrease in a sustainable society or whether the variety in human modes of life would also decrease, when what is happening is a *change* in the possible and ethically permissible range of resources available to people, rather than the *prescription* of one or a particular set of ways of life. In other words, what is happening is a change in the *means* and not the *ends* to which individuals can put these means.

That a society with less annual increases in individual disposable income or private consumption somehow necessarily represents either a decrease in overall human welfare or a decrease in the variety of possible ways of life, is to simply impose contemporary standards on the possible ways of life in the future or to use these standards by which to judge alternatives. In short, in terms of the available variety in ways of life and views of the good, a sustainable society in which the economy is not orientated to providing individualised goods and services does not necessarily stand condemned. However, it is also the case that a sustainable society would, in order for it to be sustainable, have to politically and legally regulate, monitor and, in some cases, proscribe or limit certain lifestyles or particular uses of the natural world. In doing so, a sustainable society would seek to encourage 'sustainable' lifestyles and discourage 'unsustainable ones'. Just as liberal societies seek to discourage illiberal ways of life, in the sense that liberal societies and states do not 'tolerate the intolerant', a sustainable society would not 'sustain the unsustainable'. It is to a brief outline of the issues surrounding these ethical side-constraints that we now turn.

Ethical side-constraints

One of the central differences between the ethical foundations of a 'Green, sustainable' society and a 'sustainable society' (see Figure 2.1) relates to the issue of different lifestyles and views of the good. While both Green and sustainable societies would allow for a diversity of ways of life, it is clear that a Green sustainable society would have a more limited range of lifestyles from which individuals could choose. The reason for this is that while both types of societies share a commitment to sustainability only the Green conception of a sustainable society insists on linking the latter to a commitment to ensuring symbiotic rather than parasitic relations between human and nonhuman. Thus, while it is possible to think of modes of human use/treatment of nonhuman animals which are sustainable, these may still be disallowed as undesirable (i.e. parasitic) from a Green point of view.

For example, bull-fighting, badger-baiting, fox-hunting, industrial meat production/factory farming, animal testing and vivisection, hunting and a whole host of other modes of human use of nonhumans may be seen as negligible from a sustainability perspective (in terms of their overall ecological impact or effect on biodiversity) or even seen as making a positive contribution to it – in terms of maintaining viable populations of particular species, for example. Alternatively, these forms of human-nonhuman productive relations could be seen as possibly or potentially sustainable, through technological, organisational or other forms of change and improvement. Yet, it is difficult to see how many of these forms of productive use and treatment of nonhumans (at least as currently practiced) would be compatible with the ethical principles for symbiosis. Thus, the ways of life dependent upon these and other similar forms of non-symbiotic human use or treatment of nonhumans would be illegitimate and, if supported by state/legal sanction, also illegal in a Green, sustainable society. While 'banning' or proscribing certain forms of lifestyle based on particular ab/uses of the nonhuman world sounds extremely illiberal and potentially totalitarian, and thus runs the risk of proving right those critics who accuse Green politics of being 'fascistic' and authoritarian (Allison, 1991: 170–8), it must be remembered that liberalism does not permit the flourishing of all possible ways or life. So, Green proscriptions of some ways of life fits within a long and honourable tradition of liberal humanism.

The good life in a sustainable society

For reasons of space I will focus on how Green political theory offers an alternative understanding of the good life. While Green politics is marked by its attention to the nonhuman world, it is also different from most other political theories in that it offers a less materialistic view of the good life. Although the reasons for this alternative view vary, from those who advocate it for ecological reasons as enabling humans to 'walk lighter on the earth' (Naess, 1995) to those who see materialism as spiritually void and of no intrinsic value (Dobson, 1990), to those who see the adoption of such lifestyles as the only way to achieve global distributive justice (Lee, 1993), most conceptions of Green political thought present a view of the good life for humans in which the emphasis is on enhancing the quality of life rather than 'quantitative' concerns such as wealth, income, or paid employment. In this respect, Greens are heirs to J. S. Mill and his desire to see a 'post-growth' stage of society in which people would concentrate on improving their minds, their relationships with each other (and we may add, the nonhuman world), instead of being fixated on consuming more and more goods and services.

In this way the philosophical basis of Green political theory (and so its vision of a sustainable society) thus involves the re-definition of central concepts such as 'human welfare', 'development' and ultimately the whole idea of 'progress'. According to the Green view, the various ecological and social problems humanity is presently experiencing, as derived from older forms of development or modernisation, cannot be said to constitute viable or desirable forms of human progress. What is required from the Green position is a radical (in the proper sense of this term, meaning 'to return to the roots') re-examination of some of the central tenets of modern society, particularly the dominant view of the good life as the ever-increasing private consumption of those material goods and services produced on the global market. Indeed, over the last twenty years or so the Green movement has sought to demonstrate that if one plots increases in economic growth (GNP) against a 'quality of life index' one finds that while GNP has increased in the West, the quality of life (in terms of personal security, social conviviality, environmental quality, etc.) has actually decreased (Jacobs, 1996)(see Chapter 6). Thus while Western populations may be richer than they were, the quality of life has fallen as a result of the development path pursued.

Therefore, a Green, sustainable welfare state would encourage a valuation of the enjoyment of free time over money and consumption (see Chapter 10) – in short, encourage a transformation of socially dominant views towards *being* and *doing* rather than *having* and *consuming*. The reason for this is that if, as many Greens hold, a more sustainable society necessarily requires a decrease in energy and resource use, which may led to a lessening of material standards of living (at least in the developed world), but which need not result in drastic decreases if accompanied, as suggested earlier, by egalitarian measures, then a shift to a society in which the emphasis is on 'non-material' goods such as free time makes sense and may be required in order to make it both politically attractive and thus democratically feasible.

The aim would be to create the economic conditions through which work could, where possible, be so reorganised that individuals would enjoy more of work's 'internal' goods, as opposed to seeing work purely as instrumentally valuable in terms of the income and purchasing power it generates. Such internal goods would include allowing greater opportunities for individual creativity within, responsibility for and control over working conditions. At the same time, many Greens suggest policies for enhancing the quality of life in which paid employment does not assume the predominant position it does in the conventional welfare state (Fitzpatrick, 1998). Here, overcoming the distinction between formally paid employment (in the money/cash economy) and work (which is located in non-market/money spheres) is important (see Chapter 9).

Conclusion

On the basis of what I have suggested above, there is no reason to think that the ethical basis of a sustainable society has to be radically different from some of the familiar ethical principles underpinning modern western societies. While this may strike some as odd I would like to emphasise two important issues. First, it is human action and behaviour towards the natural world and each other that is important. Secondly, the same action can be motivated by, and/or consistent with, different ethical principles or worldviews. Individuals may engage in practices which contribute to sustainability for a variety of reasons: some out of respect for the natural world, some out of concern for their descendants, some out of a sense of global justice. There is no reason to think that widely different (and perhaps even incompatible) ethical principles cannot support the same sustainable action, practice or policy.

To link the achievement of sustainable practices, behaviour and policies to the acceptance of a particular 'Green' ethical basis is not only undesirable (in terms of potentially threatening a toleration and encouragement of different, freely chosen views of the good life), but also unnecessary. Agreement around *what* should be done can be independent from *why* it should be done, in the sense that various reasons can be given for the same or similar action. In other words, the achievement of sustainability does not require that everyone in society (or even a majority) can be said to have a particular 'Green' ethical outlook or environmental consciousness.

In particular, there is no reason to accept that only an ecocentric ethical basis is compatible with, or a precondition for, a sustainable society. It is not anthropocentrism *per se* but rather a set of more familiar vices, such as arrogance, ignorance, selfishness, weakness of will and so on, which are largely to blame for unsustainability – it being these moral failings that need addressing. In conclusion, the ethical basis of a more sustainable society is already in place and quite familiar. What is required is a more active application of these principles and, in particular, their extension to environmental issues. We have the ethical principles needed for a sustainable society, what is missing is their consistent application in practice.

3
The Ideologies of Green Welfare

Mathew Humphrey

> I have claimed that ecologism is a new political ideology, worthy of attention in the new millennium alongside other more familiar ones such as liberalism, conservatism, and socialism.
>
> (Dobson, 2000a: 163)

> In its present stage of development, Green ideology still lacks the ideational complexity – irrespective of substantive messages – to rank it as equal to the main ideological families.
>
> (Freeden, 1996: 551)[1]

Introduction

The two quotes at the head of this chapter offer different perspectives on the status of Green ideology (or ecologism – I will take the two as synonymous). For Andrew Dobson ecologism constitutes a full-blooded political ideology to rank alongside the traditionally conceived ideologies of liberalism, conservatism, or socialism. For Michael Freeden, while it makes sense to treat Green political thought *as* an ideology, it is a 'thin' ideology – a constellation of ideas clustered around just a few core concepts, which lacks the ideational complexity of a 'full' ideology.[2] When Green ideology moves away from the concerns about the human/nature relationship that are central to it, it then borrows so heavily from more established ideological discourses that it is virtually indistinguishable from them. It is thus a 'thin' ideology.

This chapter will examine what is at stake in this debate in terms of the putative 'independence' of Green ideology and its relationship to other ideological forms. We shall briefly look at the perspectives on this question from a number of scholars who have characterised Green ideology recently.[3] We shall then look for evidence as to whether the 'thick' or 'thin' interpretation of Green ideology is appropriate. This involves assessing what Green theorists have to say on areas acknowledged to be outside of their *core* concerns, but nonetheless of some importance (in Freeden's terms regarding 'adjacent' concepts) – in this case human welfare, decentralisation, and liberty. Is what ecologism has to say on these subjects merely derivative of other ideologies, or does it achieve a distinctive discourse of its own? Having considered these areas we will be in a better position to consider this question of ideological independence and its import. We will also briefly discuss the question of where Green ideology 'sits' on the left/right spectrum: is it right, left, or as its proponents like to claim, neither, but instead 'out in front'?

A brief note on ideology

Both Freeden and Dobson share common ground in that they believe it makes sense to discuss these questions in terms of the existence of a Green ideology, be it thick or thin. This is not a view shared by all, and it is worth saying something about the notion of ideology at work here before proceeding to the substance of the chapter.[4] One of the people who is ambivalent about whether Green thinking constitutes an 'ideology' is the leading British Green writer and campaigner Jonathon Porritt (1984: 9):

> Having written the last two manifestos for the Green Party, I would be hard put even now to say what our ideology is. Our politics seems to be a fairly simple mixture of pragmatism and idealism, common sense and vision. If that's an ideology, it's of a rather different sort from those that dominate our lives today.[5]

Porritt is here invoking the idea that ideologies are inflexible (he describes them as 'fixed'), dogmatic frameworks for ordering political thinking which are neither pragmatic nor commonsensical. If Green ideology *is* an ideology, it is apparently not one of this sort.

This view of ideology is common to the 'endist' literature on ideology made famous by Daniel Bell (1962) in the 1960s and picked up again in a rather different form by Francis Fukuyama (1989, 1992) a few decades later. This understanding sees ideology as a phenomenon that is constituted by a dogmatic, inflexible set of ideas which mobilises masses of the population and offers them a set of epistemological filters through which information about the world can be processed in order to conform to the preconceived worldview. In the age of the technologically complex society with a well-educated population, these ideologies have lost their mobilising capacity (thus the 'end' of ideology). The classic contenders for 'ideology' on the Bell view are communism and fascism, but clearly Porritt would add the 'democratic' ideologies of liberalism, conservatism, and socialism to this list. In fact all of these are subsumed under the 'super-ideology' of industrialism (Porritt, 1984: Ch. 4). John Barry (1999a) employs the same understanding of the nature of ideology when he seeks to dismiss Green *ideology* – too simplistic and inflexible – in favour of Green *political theory*, which he believes to be more sophisticated (see Chapter 2, also).

An alternative view of ideology comes down to us from Marx and Engels.[6] This views ideology as a form of 'false consciousness' in that it consists of beliefs that are based upon how the world *appears* from a certain perspective rather than how it *is*. These appearances can however be misleading, as they are often distorted by class-based social relationships and a tendency to see socially produced phenomena as 'natural' and unchangeable. Ideology has to be 'unmasked' by true social–scientific knowledge of the world – which is one of the functions of Marx and Engel's writings. The discovery that the ruling ideas are the product of the ruling class, and founded in material conditions, is meant to be analogous to the discovery that, despite appearances to the contrary, the earth goes around the sun and not vice-versa.[7]

Neither the 'endist' nor the Marxist conceptions of ideology are employed by Dobson or Freeden, and this in part explains why they can both hold such a different view to that of Porritt and Barry. The conception of ideology employed by Dobson falls into a third, broadly 'functionalist' account of what ideology is and what it does. Although his account of ideology is somewhat sketchy, he suggests that ecologism qualifies as an ideology because it: (a) has a critique of existing social forms; (b) has a view of the good (Green) society; and (c) it suggests a mode of transition to take us from (a) to (b). Ideology thus has a function of transformative critique. This approach appears to treat ideology as a fairly ubiquitous form of conceptual political thought, although if all of (a), (b) and (c) are necessary conditions for qualification as ideology then certain forms of conservatism (for example) seem excluded. Freeden's (1996) understanding of ecologism rests on a fully fledged theory of ideology. On Freeden's morphological understanding *all* political thinking takes place in the form of structured arrangements of political concepts, and these structures are what constitute 'ideology'.

Neither Freeden's nor Dobson's conception of ideology presumes that ideological thinking is exceptionally dogmatic, nor do their approaches probe the epistemological questions of truth and falsehood addressed in the Marxist tradition.[8] Instead they seek to map and explain particular ideological constellations of conceptual structures. From within these (non-Marxist, non-endist) modes of thinking about ideology there are questions about the extent to which these structural constellations can be divided up into 'families' such as conservatism, socialism, etc., without reifying and over-schematising a complex universe of conceptual thinking. Freeden (1994: 162) stresses that

> This is perhaps the most important facet of ideological morphology: the absence of absolute boundaries which separate the features of ideological systems. Multiple instances exist of ideological hybrids that could only be described as conservative liberalism, liberal socialism, and the like ... It is empirically useless to entertain the notions of precise ideological boundaries, or of features exclusive to one ideology or the other. These are merely popular as well as scholarly conventions for simplicity's sake.

Additionally, to the extent that such divisions can sensibly be made, what divisions most appropriately map these ideological groupings? It is over this second question, in particular, that theorists of Green ideology disagree, as we shall see in the next section.

The independence of Green ideology: thick or thin?

How do we assess the ideological independence, or otherwise, of Green ideology and the implications of this for Green welfare? I will begin by examining Freeden's reasons for viewing Green ideology as 'thin', and then the arguments of Dobson and Stavrakakis as proponents of the more common 'thick' conception of ecologism. Once we have ascertained the grounds for these views we can begin our own assessment of this question, evidentially based upon the discourse of those who self-identify as 'Green', including the UK and German Green Parties.

Freeden's 'morphological' analysis of ideologies perceives the structural arrangement of political concepts in terms of 'core, adjacent, and peripheral' elements (Freeden, 1994, 1996: chs 1–3).[9] The core consists in the conceptual components that, on an empirical study of a particular ideological discourse, are most frequently found to be at the centre of theoretical concern.[10] These are usually the most highly valued concepts within an ideological discourse, and so those which must be instantiated in practice, and into whose service 'lesser' concepts are employed. The core shapes and constrains the content and structuring of these adjacent and peripheral concepts.

Freeden accepts that ecologism *has* a core, but it consists in four concepts only (the human/nature relationship, valued preservation, holism, and implementation of ecological lifestyles) and these 'are insufficient on their own to conjure up a vision or interpretation of human and social interaction or purpose' (Freeden, 1996: 527).[11] Ecologism may put the human/nature relationship centre stage, but there is nothing in the core to determine, or even guide, a Green conception of social formations or appropriate individual behaviour. The Green core does not even 'point ... in the direction of a clear *method* of reacting to such visions and interpretations' (Freeden, 1996: 527; my emphasis). Freeden points out that concepts central to the core of most progressive ideologies, such as liberty, equality, or rationality, are absent from the Green core and at most are pressed into service as adjacent concepts. This allows Freeden to explain the multivalent nature of Green political thought (the multiple variations of eco-socialism, ecofeminism, eco-anarchism, eco-fascism, Green conservatism, or Green liberalism) as lying 'not in an unwillingness to form a consensus' but 'in the inability of the conceptual core to supply a stronger constraining structure for adjacent decontestations' (Freeden, 1996: 529). This is consistent with Freeden's reading of nationalism as another 'thin' ideology, with regard to which he writes (1998: 750):

> in order to be a *distinct* ideology, the core of nationalism, and the conceptual pattern it adopts, will have to be unique to itself alone; and in order to be a *full* ideology it will need to provide a reasonably broad, if not comprehensive, range of answers to the political questions that societies generate.

Just as Freeden does not believe that nationalism satisfies these criteria, so he also does not believe that ecologism satisfies them either. To some extent this echoes Bookchin's (1994: 7) observation that ecocentric thought systems are socially naïve and thus open to co-optation by any form of conventional political discourse, from fascism to anarchism. Ultimately, for Freeden, Green ideology is no more than a young upstart on the ideological field, and it does not as yet have sufficient ideational complexity or determinate nature to be considered as a fully-fledged, independent ideology. Whether it will become so in the future is left as an open question.

How plausible is this view? Andrew Dobson and Yannis Stavrakakis both assess Green ideology to be a 'fully-fledged', independent ideology in the manner of conservatism, liberalism, and socialism (Stavrakakis does so on a reading of ideology that explicitly bases itself in part on a Freedenite approach). What lies behind the differing analyses? For Dobson (2000a: 7), an *ecological* rather than *environmental* ideology will be ecocentric, and that in itself appears to be sufficient to mark ecologism off as an independent ideology to be differentiated from other ideological families.[12] Dobson (2000a: 6) adds, however, that 'the central tenets [of an ideology] should hang together in such a way as to contribute to distinctiveness' and he believes that the core elements of ecologism do 'hang together' in a distinctive way. Dobson contrasts ecologism with liberalism, conservatism, socialism and feminism, and finds sufficient contrast with each to maintain his view that ecologism is an independent ideology. Ultimately:

> We can call it an ideology (in the functional sense) because it has, first, a description of the political and social world ... It also has a programme for political change and, crucially, it has a picture of the kind of society that ecologists think we ought to inhabit... Because the descriptive and prescriptive elements ... cannot be accommodated within other political ideologies (such as socialism) without substantially changing them, we are surely entitled to set ecologism alongside such ideologies (Dobson, 2000a: 201–2).

Brian Baxter (1999: 1) concurs with this view, noting that his own 'book is written in the conviction that Dobson is correct to suggest that a new ideology has crystallised out of the concerns ... about how human beings are related ... to the planet, which is, at least for the foreseeable future, their sole home'.

Stavrakakis (1997), on a discursive reading of Green ideology, comes at the question from a different direction but arrives at broadly similar conclusions. He views Green ideology as a new articulation of pre-existing elements around a distinctive set of 'nodal points' (somewhat akin to Freeden's 'core concepts' but expressed in the language of

Lacan, Žižek and Laclau). This analysis accepts Freeden's contention that Green ideology employs a constellation of political concepts that are, in themselves, long-standing and familiar, but because they are now structured around a distinctive set of Green nodal points, they, *in toto*, represent a new and distinctive ideology. 'Green ideological discourse is a *montage*, an articulation of various separate moments around a master-signifier, a nodal point or a family of nodal points that bind them together' (Stavrakakis, 1997: 266). In contrast to Freeden, Stavrakakis (1997: 270) tells us that 'the articulation, in Green ideology, of elements that used to be articulated in socialist ideologies does not mean that it constitutes a socialist ideology, since it is articulated around a different nodal point'. Within green ideology, these ideological elements of socialist origin are not located at the centre but at the periphery of the conceptual chain. For Stavrakakis, even concern for the human/nature relationship cannot be claimed as a new moment for Green ideology except in its elevation to a core concern (or nodal point). The latter novelty is, however, taken to be a crucial element in what distinguishes ecologism as a distinctive ideological formation.

Some evidence for independence

Is there any reason to care whether Green ideology is 'thick' or 'thin', ideologically independent or derivative of other ideologies? It can certainly make a difference to the ways in which we think about the prospects of anything that might be called a 'Green' approach to welfare, justice, equality, or any other sphere of politics outside of the core concern with ecology. If Green ideology really is 'thin', then outside of that core concern it could only exist in its 'variant' forms, and so adjacent concepts such as justice or welfare would be decontested in accord with, for example, a socialist or liberal perspective that was being grafted onto the Green core. If, however, Green ideology is both novel and in some sense complete (in that it adopts an independent position on a broad rather than narrow range of political questions), then we might well have a ideology that offers genuinely new approaches to these traditional problems that are worth investigating in their own right.

One method for testing the hypothesis that Green ideology is 'complete' (or 'full' in Freeden's terminology) is to assess Green ideological texts for the 'range of answers' they give to the political questions that societies generate. Doing this adequately would require at least a book-length study, but we can begin to address the problem by investigating Green decontestations of certain concepts that exist outside of the ecological core. In what follows I will do this for Green conceptions of social welfare, decentralisation, and liberty.

Social welfare is precisely one of those areas not considered to be in the 'core' of Green concerns on any of the above accounts. An assessment of what those who subscribe to Green ideology have to say about social welfare might give us some indication as to whether they employ a new and distinctive account of what human social welfare consists in and how it can be achieved. To illuminate the relationship between suggested social welfare policies and the underlying ideological commitments, Green social welfare policy will be mapped against its declared 'philosophical basis'.[13]

First, does the universally acknowledged core commitment of Greens to the preservation of the biosphere have significant implications for their conception of human welfare?[14] In ecological writing human welfare and biospherical maintenance are seen as mutually interdependent, in a fashion appropriate to an holistic ideology. Human welfare is taken as dependent upon a healthy biosphere. In the straightforward 'environmental' sense there are certain basic elements to human well-being (such as breathable air, potable water, or protection from the sun's ultra-violet radiation) that in turn demand the maintenance of certain environmental standards.[15] This is perhaps in principle uncontentious, although there has been considerable debate about the practical conditions necessary for its achievement and the costs/benefits of implementation. This environmentalist understanding appears to ecologists, however, as far too undemanding. It seeks to guarantee certain minimal physical human benefits, but says nothing of a deeper relationship between humanity and nonhuman nature. On the environmental view, for example, there is not necessarily any demand at all for the preservation of particular species of flora and fauna.

Thus there is a body of literature that seeks to forge much stronger links between human well-being and the natural environment. This literature makes the argument that human flourishing is dependent upon the existence of a recognisably natural environment. Examples include Edward Goldsmith's (1996; cf. Wilson, 1984) co-evolution hypothesis that human beings require the kind of environment in which they have evolved if they are to make sense of their world. Bob Goodin's (1992) argument that human beings require sense and pattern in their lives and this context is paradigmatically provided by nature. Another example lies in John O'Neill's (1993) neo-Aristotelian argument that an appreciation of the natural world is a pre-requisite of living the fully human life. All of these ecological arguments look to forge a much closer link than environmental arguments do between human flourishing and the preservation of biosphere integrity.

Conversely, such commitments entail obvious desiderata for human behaviour: in particular, the avoidance of behaviour that threatens what Greens perceive as ecosystemic stability and biodiversity. How is this to be achieved? It can, the Green Party insist (note 13 on page 207), *only* be achieved 'by popular consent, not by dictatorial or paternalistic means; but the sustainable society also requires a turning away from material preoccupations. It would be cynical and unrealistic to expect this from anyone for whom material insecurity was still possible'. Indeed if people's basic needs are not met there are 'far reaching consequences. This is expressed in anxiety, insecurity, and aggressive behaviour towards others, and exploitation of their environment'. Thus in order to achieve realisation of the core concept of sustainability/nature preservation it is deemed necessary that the material well-being of everyone is satisfied. The fundamental tool for the achievement of this is the implementation of a Basic Income scheme eventually sufficient to cover all basic needs, and within the lifetime of the first parliament of Green governance (see Chapter 8).[16] 'Welfare' is thus decontested as the satisfaction of material needs, and this is instantiated through the provision of a Basic Income scheme. Is this answer to the question of human welfare determined by the core commitments of ecologism?

'Determined' would be too strong, but I believe it can be shown that there is a conceptual pathway from the core commitment to ecological preservation to the peripheral attachment to Basic Income, and that this pathway is unique to ecologism. Other ideologies will also operate with some conception of human need, they may also articulate a belief that the state has a responsibility to meet those needs (as with socialism). However, and certainly at the current time, there are two norms at work in other ideological currents that help regulate the distribution of welfare benefits in contemporary societies (at least in the USA and UK). The first is reciprocity, which articulates a notion that welfare benefits have, in some way, to be earned through fulfilment of obligation to society (through, for example, actively seeking the chance to be economically productive). There is an even stronger commitment to a particular conception of efficiency, which, contra Basic Income, demands that most benefits should be means tested. Even if benefits are distributed of the basis of need rather than reciprocity, need has to be measured against existing income in order to ensure an appropriate distribution of the welfare benefit. Either one or the other, and sometimes both, of these principles are active in regulating the distribution of most welfare benefits. No other ideology justifies the provision of welfare benefits on the grounds that material security is essential to core *ecological* goals. The commitment that ecologically appropriate behaviour has to be brought forth in a non-coercive or paternalist manner, coupled with the belief that anti-ecological behaviour derives from material insecurity, come together in justification of Basic Income policy. Thus it seems that a fundamental plank of UK Green Party social welfare policy is constructed by the core commitments of Green ideology.[17]

Green welfare policy is also informed by a commitment to decentralisation (for references see note 13 on page 207):

> A return to smaller, more caring communities would reduce the need for both volunteers and social workers. The current role of welfare agencies would change and diminish; they would no longer carry the main responsibility for those in need.
>
> An ecological society will be made up of a self-governing communities of a variety of sizes which will regulate their own social and economic activities. Nothing should be decided at a higher level if it can be decided at a lower one. But the Green Party accepts that regional and national governments will continue to have an important role.

This maps onto the well-known ecological discourse of bioregionalism, premised upon the idea that 'naturally' defined regions of the world should exist, as far as possible, in a state of political independence and economic autarky. If a community has to survive on its own resources and live with its own pollution then it is unlikely to exceed its carrying capacity in terms of either population or economic activity.[18]

It might be suggested that this is hardly new, and that ecologism is borrowing heavily from anarchism in its proposition for autonomous, small-scale communal life, and no doubt it is.[19] There is, however, an important limitation to decentralisation in the Green picture, which gives decentralisation a particular 'Green' articulation. Once again it derives from the Green core conceptual commitments, or 'nodal points'. Decentralisation brings with it the likelihood of a co-ordination problem. Decentralisation diffuses both power and decision making, which raises the possibility that at least one self-governing region might decide to renege on Green principles. How is this to be dealt with? A prioritised commitment to decentralisation would allow this to go unchecked, even if the 'rogue' region was exporting pollution to Green regions. This, however, would not be an acceptable outcome for political ecologists, and so once again the core commitments take precedence and this is what gives the 'role' of higher levels of governance bite. Centrally imposed ecological limitations have to have priority, thus the role that the Green Party wants to retain for central government. Arne Naess (1989: 157) clearly articulates a belief that the less change there is towards more ecological behaviour through internalised norms, the more statist regulation there will have to be.[20] Jonathon Porritt (1984: 168) also notes that decentralisation 'can never do the job it's intended to do without the corresponding changes in values and attitudes. Decentralization depends as much on people accepting their personal and community responsibilities as on specific measures of devolution'.

Finally, let us look at the Green perspective on another concept declared (rightly) by Freeden to be 'outside' the core – liberty. The UK Green Party (for references see note 13 on page 207) has this to say about freedom: 'The Green Party affirms the importance of individual freedom and self-expression. We believe people should be free to make their own decision on matters which do not adversely affect others.' This appears to be a classically liberal line of freedom, in which freedom is maximised subject to the constraint of a 'harm principle' that is intended to guarantee something like an equal freedom to all. The Greens also invoke a 'harm principle' to define the legitimate grounds for intervention into the otherwise sacrosanct sphere of personal freedom, but much, of course, depends upon how that harm principle is fleshed out. Once again, the notion of harm is constructed in a distinctively Green way, such that 'individual freedom should not be exercised where that freedom depends on the exploitation or harm to any person or group in society, *or to the environment*' (emphasis added). So harm, in a very non-liberal way, includes harm to the environment, which at least potentially opens the door to significant, ecologically justified restrictions upon individual liberty.[21]

This view is also reflected in the Green position with regard to private property – almost sacrosanct in certain versions of liberal thought. 'Property laws should permit neither states nor individuals to treat their property in whatever way they choose. .... Property laws should therefore impose duties on owners as well as granting rights.' In similar vein, Jonathon Porritt (1984: 116) notes that in 'today's crowded, interdependent world ... individualistic tendencies are beginning to destroy our general interest and therefore harm us all'. Irvine and Ponton (1988: 65) propose banning advertisements for meat products and preventing the inclusion of free gifts in food goods such as cereals. While these may be trivial examples they do illustrate the inevitable tensions between a commitment to liberal freedoms, on the one hand, combined with a comprehensive conception of the good, Green society on the other. The environmental interpretation of the harm principle is a feature that clearly distinguishes it from liberalism.

The above is of course sketchy and taken from a fairly narrow informational base, nonetheless I hope it is enough to at least suggest that there probably is a distinctive Green ideological position on a range of political matters that lie 'outside' of the Green core. I would suggest that the Green core, 'thin' though it may be, is still sufficiently 'full' to significantly delimit the range of possible decontestations of political concepts that can be rendered tolerably consistent with it. Thus a substantial ideology is built up around a core that may admittedly contain few components. The important point is that the relatively small number of concepts placed within the core does not itself determine the question of ideological independence. What is also relevant is the degree to which that core shapes and constrains the full range of adjacent and peripheral concepts.

Positioning Green ideology

Thus far we have seen that Green ideology has had both 'thin' and 'thick' readings, and in particular we have seen why Freeden resists admitting ecologism to the select club of fully fledged ideologies. This turns on the alleged paucity of the 'Green core', and the lack of inherent specification of political concepts outside of that core. Having looked at ecological decontestations of social welfare, decentralisation, and liberty, I have suggested that the interpretation of these non-core concepts in Green political thought is constrained, although not determined, by the core conceptual commitments that ecological thought carries. In this, Green ideology appears to resemble the structural characteristics of ideologies that Freeden does admit to be thick rather than thin.

This also entails, of course, that a variety of configured decontestations can exist within the 'family' of Green ideologies – ecologism will have its 'eco-variant' formulations. In this ecologism may be no different to any other ideology. However, Freeden's argument to the contrary is worthy of a response.[22] Does a broad array of eco-variant ideologies of the type listed above provide us with evidence counter to my main argument? That is, does this broad range signify that Green ideology actually *is* more porous and open to co-option by rival ideologies than is the norm – thus perhaps rendering it thin rather than thick, and difficult to 'position' with respect to rival ideologies?

It would be fair to suggest that conventional wisdom places Green political thought as a broadly 'progressive left' ideology, on the same wing of the political spectrum as various forms of libertarian social-ism.[23] This is not, however, a view that has gone unchallenged and some have attempted to demonstrate to those with Green inclinations that their 'true' home lies on the political right.[24] As I suggested above, that both ends of the political spectrum could appeal to Greens for support might offer strong evidence for Freeden's thesis that Green ideology is no more than 'thin'. This is an argument that cannot be dismissed out of hand, and we need to consider how we might begin to judge the appropriateness of the 'thick' interpretation I have offered.

The first rather obvious, but nonetheless important, point to make is that other ideologies have their variants and mutations as well. The socialism of the social-democratic variety shades into egalitarian strains of liberalism, and libertarian socialism, anarchism, and some versions of Marxism can have broad areas of overlap. Some of the leading Nazi ideologues (such as the Strasser brothers) were, arguably, serious about the 'socialism' of National Socialism (Eatwell, 1995: 98). Certain liberals, such as Freidrich Hayek, with his stress on the importance of non-rational forms of communication, seem to shade off into Oakeshottian conservatism (Gray, 1987). Because, as Freeden stresses, all of the conceptual components of political language will normally have a presence within ideological discourse, and because the range of possible decontestations in any society is not infinite but limited by cultural and logical constraints, some common ground and overlapping perspectives between ideologies should be expected. Given that core conceptual commitments never completely determine the decontestations of adjacent or peripheral concepts, internal variation and differentiation around the core is also to be expected. The question is, then, whether Green ideology is any different to these other ideological formations in this regard.

Dobson (2000a) argues that ecologism is no more susceptible to co-optation than any other ideology. This is because the core commit-ment of ecologism to ecocentrism and limits to growth cannot be adopted by ideologues operating within other ideological traditions. I think the analytical understanding Dobson adopts here is broadly correct. Precisely because, as discussed above, the core conceptual cluster of Green ideology does constrain conceptual choice, other ideologies are not able to adapt themselves sufficiently in order to adopt ecological principles without thereby abandoning a significant element of their own core conceptual commitments. Whereas I believe there are good reasons to question Dobson's own conception of the core commitments of ecologism, this does not prevent one from accepting the thrust of his analysis.[25]

We can see this if we look at some of the attempts that have been made to develop ecological variants of the mainstream ideologies, such as James O'Connor's (1998) work on eco-Marxism, David Pepper's (1993) on eco-socialism, John Gray's (1993) Green conservatism and Marcel Wissenburg's (1998) construction of a Green liberalism. There seem to be two strategies open to ideologues seeking to develop new eco-variant versions of traditional ideologies. First, one can give an existing ideology an environmental façade. On this approach, the fundamental ecological political principles are not, and are not intended to be, imported into the core of the 'new' ideological construct, but rather the traditional ideology is given an environmental twist. The general rhetoric of environmental concern can quite easily be adopted without upsetting the apple cart of ideological fundamentals. The commitment to economic growth and development characteristic of most modern ideologies and questioned by Greens can be relabelled 'sustainable growth' or 'sustainable development' with no short-term requirement to do any more than pay lip-service to the reformulation. Of course the world of politics is full of unintended consequences, and in the longer term the importation of these conceptual modifiers may have real impact. But the logical possibility of mere cosmetic change is real enough, and tends to be what most ecologists consider the 'greening' of mainstream politics to consist in so far (Robinson, 1992).

The other possible strategy is to form a genuinely new eco-variant ideology that might truly be 'eco-socialist' or 'Green conservative', but which accepts the need for fundamental change with regard to the mainstream ideology as a result. The argument for the independence of Green ideology does not foreclose the possibility of developing coherent hybridised eco-variant ideologies. It *does*, however, imply that these eco-variant ideologies will possess a conceptual core that is substantively different to that of the non-eco-variants. Conservative, liberal, Green, etc. are heuristically useful labels that break up the complex universe of political thinking into workable divisions and allow us to characterise thought patterns that possess identifiable shared components and structures. None of that should blind us to the concomitant artificiality of those same divisions. Political concepts can be structured in new ways that do not fit terribly well within the existing ideological canon – otherwise new ideologies could hardly emerge. There are limits to this process (could one imagine a 'liberal fascism'?) but little more.[26]

The strategy that is probably not available (on just those grounds of consistency) is the development of an ecological variant of a mainstream ideology that leaves the conceptual core of the original ideology undisturbed. The example of the attempt to develop an ecological variant of Marxism illustrates this well. James O'Connor, Martin O'Connor and a group of associated theorists have been engaged in the project of developing an ecological Marxism over the past twenty years or more, particularly through the pages of the journal *Capitalism, Nature, Socialism* and a series of books from Guildford Press (Benton, 1996; Faber, 1998; J. O'Connor, 1998; M. O'Connor, 1994). The central conceptual innovation here is the development of a thesis of the 'second contradiction' of capitalism which leads it to consume its own resource base (see Chapter 4). This may or may not be the case,[27] but the point for us is that this concern with unsustainable resource use is articulated within ecological Marxism on purely prudential grounds. If human beings are to maintain an adequate standard of material welfare then economic development has to be on a sustainable basis. An unregulated market cannot provide this sustainability, and so an alternative method of production, distribution and exchange has to be established. This alternative offers a new foundational support for the established Marxist critique of capitalism at a time when the fortunes of conventional Marxism have flagged.

Within the discourse of eco-Marxism, however, neither of Dobson's two conditions of ecologism, and only one of Freeden's four, are met. Eco-Marxism is not ecocentric, nor does it seriously question Marxism's anthropocentric assumptions. It does not actually question the desirability of continuous economic growth, only *capitalist* economic growth. The suggestion that only the capitalist form of economic growth will be environmentally disastrous appears as an assumption interjected into the argument, rather than as an assessment of the conditions of economic growth under socialism

(Pepper, 1993). The Marxist core is retained through the non-absorption of ecological principles into that core. At the same time, another strand of the eco-Marxist project common to most attempts at developing eco-variant ideologies is a reinvention of the original canon of that ideology as historically ecological *avant la lettre* (Parsons, 1977; Foster, 2000). In this case Marx and Engels' works are reinterpreted with an ecological slant, but an inevitably contentious reading such as this is unlikely to establish a convincing set of ecological credentials without leaving certain core conceptual commitments of Marx out of the picture or at least radically underspecified. There is a vigorous literature on the extent to which Marx and Engels were 'Green', to which I have nothing to add here. Suffice to say that Marx's technological optimism as expressed in his *Critique of the Gotha Programme* has to be accounted for at the very least. My own judgement here is that the core conceptual commitment to the value of the human transformation of nature in Marx, for both individual and social development, cannot be retained without disqualifying the resultant ideological formulation from membership of the Green canon.

Dobson gives more examples of such conceptual incompatibilities, which I would broadly endorse and which I believe do help to demonstrate that ecologism is an ideology in its own right. As for where it belongs on an 'ideological spectrum' we come back to the issue with which this section started. Ecologism has been claimed by both left and right, and Greens themselves like to claim to be 'neither left nor right, but out in front'. Part of the problem here is that the dimension in which the terms of the left–right spectrum are discussed is normally that of income distribution, with egalitarians to the left and non-egalitarians to the right. Ecologism may be, as I have argued, an ideology of comparative stature to the traditional ideologies, but this does not entail that it will necessarily sit easily on the left-right spectrum, where a ranking for environmental concern has never been part of the left–right equation. I would suggest that in ecological discourse a commitment to a redistributive position on the left-right spectrum can only ever be instrumental. That said, it is crucial to bear in mind that an instrumental reason for doing x (in this case redistributing income) can be a very strong reason indeed where a, (in this case sustainability) the intrinsically valued object, is valued highly, and x is considered crucial in realising a. By contrast, intrinsic reasons for performing a duty might be quite weak.

To refer back to the UK Green Party, their position on welfare policy has an egalitarian flavour in the shape of Basic Income (egalitarian in the sense that it at least guarantees all the same minimum income, even if it allows variations above that minimum to develop). As noted in Chapter 8, however, this is regarded by the UK Greens as the most appropriate means for achieving their highly valued core aspiration of the ecological society. On the grounds that people whose material security is catered for do not have an incentive to be ecologically destructive, we have a Green justification for unconditional Basic Income. This policy does place UK Greens towards the left end of the spectrum, in the general proximity of socialism and egalitarian liberalism. However (as with Bob Goodin's characterisation of Green political theory)[28] one can reasonably question the strength of this 'left' redistributive commitment. On a Goodin-type consequentialist reading of Green ideology, that commitment only holds true as long as Greens have good reason to believe that this really is the most efficacious means to bring about valued Green ends. If Greens have good reason to believe in another form of agency – that a highly inegalitarian society could, for some reason, best deliver Green outcomes – then it would be entirely appropriate for Greens to be radical inegalitarians.

Much depends, then, on the degree to which one reads an intrinsic connection into the relationship between Green outcomes and egalitarian means. A Green ideologue may well accept that egalitarian-ism is indeed a means, but seek to break Goodin's radical means-end distinction by claiming the causal relationship between egalitarianism and Green outcomes is so strong that an inegalitarian Green ideology becomes an oxymoron.[29] If this latter case can be made then again we have good reason to adopt a 'thick' interpretation of Green ideology.

Conclusion

I have argued in this chapter that Green ideology stands as a full fledged, independent ideology that, although its 'core' might be rather thin, nonetheless has a distinctively Green approach to issues across the field of politics. The ideological core of ecologism is sufficient to shape and constrain the decontestations and structural arrangement of adjacent and peripheral concepts. Furthermore we have seen in the last section that the existence of particular policy positions can plausibly be seen as being strongly guided by core conceptual commitments. There are specifically 'Green' positions on liberty, decentralisation, and human welfare, to take only the case studies from this chapter. I believe that there are distinctive Green positions on a broad range of other issues as well (although demonstrating this would require a considerably longer and more systematic study). None of this entails that Green ideology employs a whole new political vocabulary, far from it. It employs a language of politics that anybody reasonably acquainted with conventional western political discourse can understand, although it has formulated a few neologisms as well, such as the commitment to ecocentrism that Dobson places in the core. From the perspective of this chapter there is a Green interpretation of social welfare. Although the policy recommendations (at the ideological periphery) may not be specific to Greens, the path of justification for this policy (the derivation of policy from the ideological core) is unique to this ideology.

4
Green Democracy and Ecosocial Welfare

Tony Fitzpatrick

Introduction

The purpose of this chapter is to sketch the outline of a theory of ecosocial welfare (also, Fitzpatrick with Caldwell, 2001). This theory lies at the intersection of three lines of inquiry. The first line concerns the meaning of, and the prospects for, a Green democracy. Here, I contend that any Green democracy must be rooted in the values and goals of liberal democracy, both for reasons of principle and of expediency, but that only a 'radical liberal democracy' can realise *ecological* values and goals. The second inquiry concerns the relationship between democracy and the welfare state. Is this one of mutual support, conflict, or a bit of both? A later section will review the main arguments in this respect. The final line of inquiry concerns Green critiques of the welfare state, critiques that Michael Cahill and myself have already outlined in Chapter 1 and which are dealt with throughout this book. Having laid the groundwork, the final substantive section will then discuss the main features of ecosocial welfare: a commitment to radical reformism, to 'discursive reflexivity' and to the democratisation of the ownership and control of both social and natural resources. Before proceeding, however, we need to identify the ideological sources of ecosocial welfare and this is the task of the next section.

Post-Marxist liberalism, anyone?

There is only space here to deal with three of ecosocial welfare's ideological sources. The most important omission is feminism, though significant elements of this appear within Fitzpatrick (1998).

Anarchism is also omitted, but this is less of a hardship given the anarchist tendency to somehow both overestimate the importance of politics (by treating the economic as an effect of the political) and to underestimate that importance (by proposing to abolish the state) (Carter, 1999).

The title of this section is therefore deliberately playful as there is little need to essentialise a debate whose strengths are its openness and eclecticism. Even so, I wish to underscore the importance of each of the following.

Liberal democracy

The liberal democratic state is now the accepted model of political organisation in more areas of the world than ever before because the complexity and inter-relatedness of that world makes pluralism – the dispersal of power across multiple centres of interest and representation – more suitable than any of its alternatives. In truth, actual liberal democracies embody a form of 'restricted pluralism' since some centres are obviously more powerful than others. Nevertheless, liberal democracy aims at the ideal of a level playing-field upon which each player can mobilise, contest and/or co-operate with others in order to promote their values, interests and agendas. Some individuals and groups may be more powerful than others at any one time, but no-one on this field should have an unfair advantage.

The liberal democratic state is supposed to underwrite this equality of political conditions: by acting as a neutral umpire and as a defender of minority rights, the state prevents majorities from wielding despotic tyranny over others. Such neutrality implies a series of checks and balances to ensure that those who are winning the political game at any one time are still open to challenge within the political arena. In reality, liberal democratic states are rarely neutral in that they inevitably reflect the religious, historical, ethical and national contexts within which they developed. Even so, liberal neutrality remains the ideal and, as such, there are basically two models to which this state may conform. First, there is the minimalist model where the playing-field consists of free and unregulated market relations and a residual state guarantees private property and contractual exchange (Nozick, 1974). Secondly, there is the managerial model where the state effects the fair and just distribution of social goods and benefits that the free market violates (Rawls, 1972). The welfare state derives from the latter model.

We shall be examining the debate concerning Green liberal democracy in the next section. It is worth pointing out here, though, that many within the ecological movement have supported a radical re-organisation of liberal democratic precepts and institutions. Sagoff (1988) believes that liberal democracy can be preserved only if our altruistic and long-term preferences as citizens are given priority over our selfish and short-term preferences as consumers. Ophuls and Boyan (1992), though, believe that this trick cannot be achieved without going beyond the current democratic system, where the political marketplace is dominated by the key actors within the economic marketplace. Two approaches to social policy therefore suggest themselves: if Sagoff, Ophuls and Boyan are correct then a new approach to social welfare systems has to be devised, an approach that, whilst building upon the existing system, may take us in directions currently difficult to imagine and conceive; but if they are wrong (as Wissenburg believes, see next section) then social policies become a means of reconciling our Green aspirations to the existing institutional order. In its present formulation, the theory of ecosocial welfare is agnostic as to which of these approaches to Green social policy is superior.

Marxism

Marxists contend that the state in a capitalist society is biased towards the needs and interests of capital. Far from being a neutral umpire, the state is that which (1) conspires to ensure that the playing-field of market capitalism maintains a steep gradient, and (2) tries to convince us that the weaker players are responsible for their own misfortune. In short, the state is concerned with securing the conditions for the accumulation of capital and with the moral legitimation of social injustice and exploitation (J. O'Connor, 1973; Habermas, 1975). However, Marxists are then torn between two elaborations of this basic interpretation.

The 'functional model' treats the state as an unassailable citadel that both socialises us into the performance of capitalist imperatives and suppresses dissent (e.g. Althusser, 1969); the 'conflict model' treats the state as a site of struggle and therefore as a potential means of securing social justice *within* capitalism and of transforming society *beyond* it (e.g. Gramsci, 1971). While these models once generated a great deal of argument, many are now content to draw upon both. Offe's (1984) famous observation that the welfare state is both necessary for, and a threat to, capitalism expresses this compromise succinctly.

Eco-Marxists have taken the debate one step further (Benton, 1996). In its classical formulation (Marx, 1977: 388–91) major historical transformations result when the contradiction between productive forces and production relations becomes irreparable, and the state is overthrown when one system of property ownership must be replaced by another. According to James O'Connor (1998), though, this captures only the *first* contradiction of capitalism. The *second* contradiction is that between the forces/relations on the one side and the 'conditions of production' on the other. In essence, capitalism is self-destructive since the profits and the over-production that it feeds upon must eventually drain away the natural resources upon which any society ultimately depends. Capitalism has survived so far by producing and distributing an immense affluence, but this has only had the effect if externalising and displacing its first contradiction onto the second by which the natural conditions of its own future survival are undermined.

And what of the state? Quite simply, the state mediates between nature and capital by regulating capitals' access to, use of and exit from production conditions (J. O'Connor, 1998: 148–55). The capitalist state brakes and steers the depletion of resources and the pollution of the environment, though it cannot change the overall direction of society. But just as many Marxists have argued that the state *can* be a force for social justice and progress, so eco-Marxists insist that it *can* be a force for environmental justice and sustainability. If capitalism is characterised by two contradictions then so is the state: the first emerges through the systemic conflict between accumulation and legitimation; the second between the desire for accumulation and the need to both observe and preserve natural limits.[1] Therefore, just as the first contradiction can be exploited by those class movements committed to socialism, so the second can be exploited by those who are or should be committed to eco-socialism, i.e. all radical class and social movements. Capitalist social and property relations cannot create sustainable conditions of production to the extent that is necessary because this will require a degree of bottom-up, democratic planning that is inconsistent with even the most forward-looking proposals for Green capitalism (J. O'Connor, 1998: 246–7).

One implication of this is that if the state possesses a second contradiction then so does the *welfare* state. The first contradiction is that the welfare state is *both* radical and reactionary since it exhibits progressive *and* regressive features, corresponding to the imperatives of legitimation and accumulation, respectively; the second contradiction is that *the welfare state helps to regulate the depletion of the very resources, and the degradation of the very environment, upon which it itself depends.*

One problem with eco-Marxism (see Chapter 3 also) is that we are left with no clear prescriptive reforms with which to work. O'Connor recommends the democratisation of the state, but does this imply anything more than liberal proposals for strengthened local government, citizens' juries and parliaments, etc? Are we talking about the kind of direct democracy that some Marxists have long favoured? In which case, how are the oft-noted problems of direct democracy to be surmounted? Or is some form of associative democracy more appropriate (Hirst, 1994; Cohen and Rogers, 1995)? If so, then does eco-Marxism really have anything distinctive to offer such proposals?

Post-structuralism

Some ecological theorists have found it more intellectually profitable to draw upon post-structuralism and, more specifically, a Foucauldian analysis (e.g. Darier, 1999). This is because post-structuralism makes two potential contributions to ecological thought.

First, it usefully directs our attention away from traditional conceptions of power and sovereignty (Foucault, 1975, 1977). Both liberals and Marxists are guilty of identifying power as a unitary, centralised, repressive and centripetal force; but according to Foucault, power is productive and decentralised, present at every node and relation of the social network as the capillaries of the social body. Therefore, we should abandon familiar theories of the state and of political sovereignty which, by focusing upon the 'centre', construct power at a far-off, impersonal distance (indeed, this construction is itself an effect of power). We should talk instead of governance, or the 'conduct of conduct', i.e. the way in which we are made as subjects (both as subjectivities and as the subjects of power). Two points follow from this. First, we are all of us 'deviants' organising and organised around a series of norms. Governance is therefore discursive and so is not captured by the traditional sociological distinction between agency ('we govern') and structure ('we are governed'): disciplinary governance consists of norms discursively reflecting back on themselves through practice and habituation. Secondly, freedom and governance are inseparable, each depends upon the other (Dean, 1999; Rose, 1999a). Modern society consists of the simultaneous empowerment and disempowerment of both individual and collective bodies. In short, post-structuralists insist that ecologists cannot afford to limit their analysis to traditional theories of state power.

The second contribution is the post-structuralist injunction to avoid deterministic interpretations of nature and of environmental crises. One such interpretation contends that Green values have developed in response to an increasing consciousness of environmental problems and hazards. However, this common-sense view overlooks the extent to which the environment is the means for the 'carceral' management of bodies and populations. This does not simply mean that humans create the crises that they then see themselves as responding to, but that the signifier 'crisis' is itself a discursively constructed meaning. Ecologism is therefore called upon to question its self-image as a naturalistic philosophy, unpolluted by existing ideologies; indeed, ecologists are urged to excavate the negativities that lie at the heart of their critique, e.g. the possible racism and occidentalism that inspire the literature on population control. Nevertheless, eco-governance cannot imply a transcendence, merely an enhanced awareness of the power relations within which the human/nature matrix is implicated. The 'human', nonhuman nature and the 'social' are bound together discursively and cannot be separated out according to some normative blueprint, whether liberal or Marxist.

And this may be the problem with post-structuralism. By eschewing any notion of a 'regulatory ideal' it enjoins us to do what we already do, only with greater understanding, and deprives us of a practical radicalism where society *a* can be preferred to society *b* even at the risk of re-masking the power that post-structuralists labour to unmask (though with what end in mind is not always clear). More specifically, although Foucault (1984: 51–75) is correct to argue that the state cannot and does not occupy the entire territory of power, he perhaps underestimates the extent to which power retains a centralised and visible demeanour: for instance, if capital is regarded as nothing more than another discursive face of power then we seem to have done nothing more than return to the playing-field pluralism of liberalism (cf. Hindess, 1996). For instance, it is not clear whether Torfing's (1999: 225–41) discourse analysis has anything to say about the welfare state that is particularly distinctive and so, by extension, whether it has anything to contribute to the greening of social policy.

*   *   *

Liberal democratic, Marxist and post-structuralist ideas all, then, have something to contribute to an ecosocial welfare theory. Liberal democracy is an indispensable starting-point due to its basic principles (of autonomy, especially), its adaptability and its popularity; though one that can and must be disengaged from its origin as a philosophical system for the defence of male property-owners. Marxism continues to remind us that liberal democratic societies achieve a limited form of freedom in the absence of any democratic ownership and control of resources. Post-structuralism warns against any attempt to force our ideas into intellectual totalities, and it draws attention to the inescapable presence of disciplinary norms and strategies of governance. Having laid the groundwork, then, we can proceed towards a theory of ecosocial welfare by first investigating the debate concerning Green democracy.

Green democracy

Could there be such a thing as a Green democracy and, if so, what form should it take (Doherty and de Geus, 1996)? According to Robert Goodin (1992: 116; cf. B. Barry, 1995: 149–51) the problem is as follows. Democracy refers to a set of procedures the outcome of which cannot be known until those procedures have actually been carried out. If we polled a room full of 100 people as to whether they support the implementation of a proposed Green reform then the procedure (the voting) takes place prior to the outcome (the majority's approval or rejection of the proposal). The trouble is that ecologism is extremely consequentialist, i.e. it focuses upon the outcome rather than the means. If there is no guarantee that electorates will support Green proposals then what should Greens do? They either have to ignore the principles of democratic proceduralism, which implies adopting authoritarian and, perhaps, terroristic strategies, or they have to support principles which may require them to accept the non-implementation of Green reforms wherever the majority does not vote in their favour. Green democracy is therefore impaled on the same dilemma that all political radicalisms face: either a Leninist approach needs to be taken, or else democracy needs to be prioritised above one's cherished beliefs and values.

However, it could be argued that the dilemma is not this severe after all on the grounds that there is no such thing as a pure proceduralism. Voters do not exist *ex nihilo*, in a social and ideological vacuum, but are embedded in a particular culture into which they have been socialised. Equally, political and economic elites are able to shape the flow of ideological opinion, whether deliberately or not, so that the votes of the majority may do nothing more than reflect the interests of the powerful. Finally, it can be argued that procedures and their consequential implications change depending upon whether the system is one of representative, direct or participatory democracy. The civic competence of actors in a participatory democracy may well be higher than that of existing voters so that the procedures of such a democracy would be more conducive to Green reforms than that of a representative one. If, in other words, procedures are always skewed in one direction or another – if all democratic systems are ends-oriented – then there is nothing self-contradictory in ecologists trying to skew them in Green directions away from the non-Green directions that currently prevail.[2]

What might this imply? At the theoretic level it could imply the enlargement of the moral community. Both nonhuman species and future generations of humans and nonhumans might be regarded as the objects of moral and political consideration, giving the democratic system less of an anthropocentric facade. Just as prior democratic revolutions have shifted the centre of political gravity away from male property-owners, so a further revolution could shift it away from its overwhelming domination by the interests of the present generation of just one species, ours. For instance, Dobson (1996) suggests that legislatures can be revised to incorporate specific representatives of future generations and this proposal could be extended to cover nonhuman species also. In addition, some have argued for a category of environmental rights (Benton, 1993; Saward, 1996): the rights of humans to live in and enjoy a clean, sustainable ecosystem and the rights of animals to be free from cruelty. If such a category of rights is deemed acceptable then democratic institutions must represent and preserve them no less than civil, political and social rights. Committed democrats therefore have an incentive to support ecologism and ecologists have an incentive to support democracy. So, there is a *prima facie* reason for believing that, yes, there can be such a thing as Green democracy: for just as democracy is always aimed at particular ends, so ecologist values can be incorporated within majoritarian procedures.

However, what form might this Green democracy actually take? There are five basic possibilities. First, we might favour a Green market liberalism where environmental costs are factored into the price mechanism. Secondly, we might favour a Green social democracy where redistributive and managerial means are employed according to the requirements of social *and* environmental justice. Thirdly, we might favour ecosocialism where the popular control and ownership of the economy is extended to the environmental preconditions of economic activity. Fourth, we might favour some kind of eco-centralisation within which there is only a limited form of democratic representation and participation. Finally, we might favour an eco-anarchism where the state is abolished and replaced by 'horizontal' networks of egalitarian, democratic and autonomous communities.[3] I shall rule out the first of these options on the grounds that markets must always be regulated if they are contribute to socially desirable ends; the fourth option is also dismissed because it harks back to the survivalist discourse of the 1970s which is now both scientifically and politically outdated (Dryzek, 1997); and the final option is rejected for the reason noted briefly at the beginning of this chapter. This leaves us with two alternatives, then: either a Green social democracy or an ecosocialism. In order to decide between them we need to revisit the debate concerning liberal democracy.

The most persuasive recent case for simply greening the status quo has been made by Marcel Wissenburg (1998). Wissenburg does not argue for a Green social democracy himself (cf. Jacobs, 1996, 1999), but he does offer a theoretical defence of a Green liberalism that would need to underpin any Green social democracy. On the plus side, he offers a necessary corrective to those who are casually dismissive of liberal democracy on the grounds that it is too anthropocentric a system (Wissenburg, 1998: 103–6). Nevertheless, crucial parts of his argument rest upon some shaky foundations. Basically, Wissenburg offers too restrictive and limited a definition of liberal democracy, allowing him to polarise the relevant debates and overlook various conceptual distinctions that are subtler than those with which he works. There are three specific problems with his approach.

First, Wissenburg (1998: 63–5) makes a crude distinction between the piecemeal engineering of ecological modernisation and the more radical engineering of 'ecological utopianism'. Setting up the terms of the debate in this simplistic fashion loads the dice against those who believe that piecemeal reforms must take utopian goals as their regulatory and long-term ideals if we are to achieve not so much utopia as what James Meade (1993) refers to as 'agathotopia', i.e. the best possible world rather than the perfect world. Although probably correct in his observation that 'liberalism is compatible with reformist policies only', Wissenburg (1998: 200) then identifies reformism with the 'least socially disruptive policy that empirical circumstances allow' and so social democracy is the furthest left that he allows us to go. In short, he is equating reformism with 'conservative reformism' and airbrushing from the picture the possibility of a 'radical reformism' that derives from a reflective equilibrium between principles of justice and the practicalities of given circumstances. In other words, Wissenburg neglects the possibility of what Christoff (1996) calls a 'strong ecological modernisation' which would be more discursive and less technocratic than the 'weak ecological modernisation' of current environmental policy-making.

This relates to a second key problem. Wissenburg (1998: 191–4) treats liberal democracy as a system for the aggregation of preferences rather than as that which shapes the norms, expectations and very identities of preference-holders in the first place. Not only does this underestimate the extent to which liberal democracy 'governs the self' – the insight provided by post-structuralists and many Marxists – but it over-states the extent to which interference in the formation of preferences necessarily leads us to 'far less attractive political systems'. We face a policy choice not between interference and non-interference in individuals' preferences but between (a) reflexive interference, and (b) nautonomous interference: the latter deriving from the logic of economic and political systems, the former referring to the inter-relational deliberations of autonomous citizens – which is precisely why some insist that liberal democracy must subsume into a discursive democracy (Fitzpatrick, 2002). If preferences are constructed rather than given then we have reason to believe that free choices can be made more Green whilst still remaining free – just as consumerism also represents a type of free choice, if a highly limited type according to Greens.[4] Engaging with this problem of preference-formation means discussing what, later on, I shall refer to as 'discursive reflexivity'.

Finally, Wissenburg's economy (1998: 212–19) consists of individuals, of producers and consumers, but not of classes, strata, groups and structures. So although it is not clear what his defence of 'economic liberalism' amounts to – by this he does not necessarily mean *laissez-faire* – his methodological individualism suggests a preference for deregulatory exchange and trade. Wissenburg (1998: 213) distinguishes between markets and market preferences as if the former have got nothing to do with the latter:

> if 'the' free market is to be blamed for environmental problems, it cannot be blamed because of its being a *free* market; it is because there are consumers and producers who are looking for a market – any kind of market.

While in an abstract sense it is true that consumers *could* choose to pay higher prices for Green goods and so alter the outcomes of free market exchange (as some choose to purchase more expensive organic foods) there is a significant constraint on the widespread adoption of such preferences. The immense income inequalities that accompany free market capitalism both inhibit the spending patterns of the poorest and motivate the non-poor to spend their disposable income on (1) positional goods, and (2) savings and investment, to insure against the ever-present possibility of downward mobility. Wissenburg (1998: 218) acknowledges that government regulation can ameliorate environmental cost evasion and corporate short-termism but he neglects to say how this can be done given the power of global capital – indeed, he seems dismissive of arguments that draw attention to the latter. So whereas Green capitalists can certainly find a niche or two (witness the continued success of The Body Shop), these are unlikely to expand into anything more without effective state regulation. Textbook economists can construct perfect models of Green capitalism, just as market equilibrium can be demonstrated mathematically, but, in the real world, how can we transform the economy from here to there without more extensive state and inter-state action to alter the very preferences (of producers and consumers), geared to self-interest and short-termism, which Wissenburg demands that we leave alone?

Wissenburg's defence of a Green liberal democracy therefore runs up against certain problems, the resolution of which allows a case for a more radical political economy to be made – as Sagoff *et al.* want to suggest. It is in this context that ecosocialism becomes relevant. Rather than pursuing this here, though, my aim is to design and defend a theory of ecosocial welfare, a theory that lies between Green social democracy and eco-socialism. Before doing so, however, we must first review the second line of inquiry mentioned in the introduction.

Democracy and the welfare state

Having argued that there can be such a thing as Green democracy and that liberal democracy represents a convenient and principled *starting-point*, we can start to introduce social policy into the frame. First, what is the relationship between democracy and the welfare state?

The previous section has offered a deliberately upbeat assessment of the potentially mutual benefits for ecology and democracy. It may well be, however, that attempts to Green liberal democracy will miscarry and electorates will democratically decide not to avoid environmental degradation, whether due to ignorance, deception or sheer perversity (see Chapter 2, however). Democracy is no panacea, therefore. That democracy may enhance but cannot guarantee higher levels of social welfare is a lesson that social policy commentators have long appreciated and it is worth pausing briefly to appreciate why.

The welfare state has often been accused of being too bureaucratic and dependent upon the unaccountable opinions of faceless administrators and experts. The further back in time we go the less this accusation surfaces, though. Indeed, the earliest architects of the welfare state made a strict distinction between the voluntary and mutual aid sectors, where the efforts of non-experts would be required, and the state sector, where services would be run *for* people but not, by and large, *by* them. The idea that those who received state services could have a considerable voice in their design and operation did not gain a firm foothold until the 1970s, by which time it was too late as the consumerist reforms of the Right were gaining in popularity. Nevertheless, the idea that the welfare state requires a democratisation has remained popular on the Left ever since, and deservedly so (Fitzpatrick, 2002). However, it would be counterproductive to over-emphasise the role that further *political* democratisation could play, as Claus Offe (1996) makes clear.

Offe observes that there are four ways of conceptualising the interaction between political democracy and the welfare state. First, we can think of the former as supporting the latter. For instance, the achievement of a universal franchise both reflected the growing strength of the working-class and, in turn, encouraged the labour movement in its social and political struggles against the dominance of the market and of employers. Secondly, the welfare state can be supportive of political democracy, e.g. by humanising liberal capitalism and so integrating most people into the political mainstream by reducing the attractiveness of ideological extremism. Thirdly, however, the welfare state may possess anti-democratic implications by facilitating the kind of top-down corporatism that seems less than conducive to political accountability and accompanies impersonal forms of bureaucratic power. Finally, political democracy may threaten the welfare state under certain circumstances; for instance, by enabling the formation of anti-welfare coalitions, e.g. in the form of taxpayer revolts, that override considerations of justice, equity and a fair redistribution of resources. Since democracy, at least in its modern liberal variants, appeals to individualistic self-interest much more than to the common good (unless expressed in crude nationalistic terms) then the collective goods of the welfare system are constantly on the defensive. The lesson might well be that whereas democracy can assist in the formation of welfare systems when the power of labour is increasing relative to that of capital, when there is an anti-labour backlash (as there has been for several decades now) democracy may militate against further expansion and may encourage retrenchment.

Short of falling back on authoritarian solutions we are left with the conclusion that if there are inherent limits to the justice-enhancing potential of political democracy (environmental as well as social justice) then there are basically two things we can do. First, we can either resign ourselves to these limits and set about engaging the right *and* the anti-environmental left in a form of political trench warfare where all sides expend a great deal of time and energy in capturing a few yards from their enemies. This implies that any gains in social and environmental justice are likely to be vulnerable to attack and may short-lived. Secondly, we can revive arguments for radical forms of both political and economic democracy. The problem, in short, may not be democracy *per se*, but the kind of political democracy that is detached from its citizens and from the collective ownership and control of both social and natural resources. The theory of ecosocial welfare is an attempt to think through the implications of this second alternative.

Ecosocial welfare

Here is the argument so far. In order to secure a future of both social and environmental justice it is necessary to strengthen the theoretical and practical (policy-oriented) links between democracy and ecologism. However, limiting ourselves to a Green liberal and social democracy may be insufficient. The subject of social policy suggests that democracy and social justice are only mutually enhancing when capital's power is balanced out against the countervailing imperatives of oppositional movements. More than ever, then, we need to think beyond liberal democracy to some form of eco-socialism. The theory of ecosocial welfare represents a compromise between Green social democracy and eco-socialism, a means of effecting the transition from the former to the latter (Fitzpatrick with Caldwell, 2001) by discursively constructing a political position around which contemporary oppositional movements could be invited to mobilise. I have outlined the ideological sources of this theory and, in a critique of Wissenburg, I alluded to three elements of this theory: radical reformism, discursive reflexivity, Green economic democracy. In this final section we shall review each of these in turn.

Radical reformism

A politics of radical reformism stretches across several camps. For instance, it is concerned with the practicable and the 'do-able' but also with the utopian and the visionary; it looks to the short-term as well as the long-term; to the abstract and hypothetical as well as to the concrete and down-to-earth; to the macro-engineering of the state and to the micro-engineering of civic experimentation. In short, the policies of radical reformism are not developed as blueprints that then require translation into practical reality, nor are they about following the status quo; instead, radical reformism implies the attempt to connect revolutionary ends with pragmatic means. It is not so much a set of prescriptive policies – although these can be obviously be proposed – as a project that is not afraid of being simultaneously conservative and extremist.

The key lies not in overturning existing systems but in identifying their emancipatory potential and taking that dynamic logic to its radical conclusions. For instance, Basic Income is a resiliant idea (see Chapter 8) because it offers a minimum income guarantee without the Right-wing insistence upon means-testing and it offers a form of social insurance without the holes and inadequate coverage of contribution-based schemes (Fitzpatrick, 1999a). Existing social policies may well be janus-faced, concerned with social control *and* social justice to varying degrees, but a post-contradictory state of affairs can be imagined if we strategically confront politicians, taxpayers and clients with the contradictions of the existing system and offer possible paths to their resolution. Radical reformism therefore proceeds by identifying and exploiting the systemic infirmities of market capitalism.

Yet at the same time as revolutionising the logic of conservative means so radical reformism also seeks a de-radicalisation of utopian ends. This means aiming at an agathotopia. Agathotopia represents a necessary compromise between realism and idealism: without the latter the former has little purpose or direction, but without the former the latter has little meaning and significance. For example, this may mean introducing into the ideal of renewability an acceptance that the deple-tion of renewables and non-renewables can be slowed but perhaps never eliminated (Fitzpatrick, 2001a). If this compromise still allows the projection of human civilisation several millennia into the future (and how much longer can we realistically plan for anyway?), until such time as the future can be left to apply reforms currently unimag-inable, then existing forms of socioeconomic activity can be adapted to this long-term but finite time-horizon.

In less abstract terms, radical reformism involves the convergence of certain theoretical priorities with the experimentations and innovations implemented on an often hesitant and fragmented basis by activists and citizens within the third sector – or what I prefer to call the 'fifth sector' (Fitzpatrick, 1999b). This convergence occurs through the understanding that each is the condition for the other. For instance, the main priority that has emerged from within ecosocialism concerns the redistribution of domestic work and paid employment so that the former is less the province of women and the latter less the province of men (Little, 1998). In addition, such redistribution may enable an overall reduction in the amount of paid employment that we collectively perform, freeing up time for people to participate in fifth sector activities (leaving open the question as to whether incentives and disincentives are required if widespread participation is to be encouraged). Such activities presently take the form of a range of civic experiments, many of which localise and collectivise economic exchange. One such experiment revolves around local currency schemes (Offe and Heinze, 1992) and both this and work/employment redistribution are discussed in Chapters 9 and 10 respectively.

In conclusion, radical reformism implies a reflexive project whereby conservative means are revolutionised and utopian ends are de-radicalised through the mutual convergence of theoretical priorities and civic experimentations. The term 'ecosocial welfare' is therefore meant to capture two arguments. First, that ecosocialism is a goal at which we should aim but which we *may* never achieve. It could very well be that a Green social democracy is the most we can ever hope to achieve, but this conclusion should be the result of the political imagination engaging with civic experimentation and not, as for most contemporary social democrats, a *fait accompli* that stifles debate and more radical visions of society. Secondly, ecosocial welfare expresses the intuition that both the concept of welfare and the state provision of welfare are likely to remain central to any strategy of reforming our productivist societies. The welfare state has usually failed to meet the expectations that many have had for it, but rather than assist the right in its dismantlement the Green movement should surely engage with social policy in the hope that a declining environment may provide an impetus for radical reform.

Discursive reflexivity

There are two aspects to this. First, we need to follow Torgerson's (1999) Arendtian distintion between functional, constitutive and performative types of politics. Functional politics corresponds to Arendt's (1958) category of 'labor' (sic), that is, of instrumental and rationalistic economic activity. The ends of functional politics are extrinsic to it and, in Green terms, takes the form of weak ecological modernisation, i.e. a reform strategy whereby ecological imperatives are adapted to established goals such as the expansion of GDP growth. Functional policy-making therefore tends to be administrative and technocratic, concerned with mechanisms rather than morals. Constitutive politics derives from Arendt's category of 'work' and is also an instrumental politics of extrinsic ends: the maintenance of the existing socioeconomic system. However, constitutive politics is also concerned with the cultural ground of civilisation and civilised practice and so is open to a greater degree of radicalisation than that of functional politics. This, then, represents the sphere of strong, or discursive, ecological modernisation (Hajer, 1995: 280–1): one where ends and means are open to at least a limited form of reciprocal negotiation and redefinition, as suggested in the previous section. So, if functional politics treats social reform as a 'hard' technocratic enterprise, for constitutive politics society is 'softer', more malleable and plastic in its values and cultural identifications.

By contrast, the ends of a performative politics are intrinsic to political action itself: they are neither given nor stable, but emerge repeatedly in different motifs, in and through the exercise of political practice and discourse. Performative politics is the modern equivalent to Aristotelian *phronesis*: that which connects worldly affairs to its originary roots in ontology, morality and contemplative philosophy. It is the source of individuals' mutual recognition of each other as citizens who share the same fate. There is, then, no such thing as performative policy-making; rather, the performative public sphere is the space to which we ascend once we have laid the functional and constitutive foundations. The performative public sphere therefore consists of self-referential debate whose object is the public and private good.

To introduce the concept of the good is not necessarily to get caught in the interminable liberal/communitarian discussion. For instance, one of the least remarked upon aspects of Macintyre's (1982) work concerns his notion of internal practice. To simplify, internal practices are those performed for their own sake whereas external practices are performed for reasons and objectives unrelated to the essential characteristics of the practice itself. For Macintyre, virtue (the constant attempt to realise the good) is such an internal practice: if a morally-justifiable action is performed for money then it is, by definition, not virtuous, however beneficial its consequences. It is the 'internality' of virtue that performative politics takes as its rationale and subject-matter:

> the good life for man is the life spent in seeking for the good life for man, and the virtues necessary for the seeking are those which will enable us to understand what more and what else the good life for man is (Macintyre, 1982: 219).

Adapting this slightly we can claim the following. The individual who experiences a high degree of welfare is he/she who has the time, the resources and the social capital to debate, with others, both the subjective and objective meanings of welfare, and a high welfare society is that which recognises the non-instrumental value of, and so facilitates, such debate (Fitzpatrick, 2002).

The ecosocial welfare model therefore stretches along a line of reflexivity between each form of politics. At one extreme lies the nitty-gritty of income maintenance, health care, shelter, etc.; at the other lies an equation of welfare with 'the good'; and in between lies the radical reformism that contruct the walls of civility that actually shape and enable the public sphere. Ecosocial welfare incorporates, but is not reducible to, social policies and systems of state welfare; at the same time, the concept of ecosocial welfare suggests that questions of the good must be attached to ecological issues and themes, as a batch of moral philosophers have begun to do. Living the good life, a life of welfare, increasingly refers to the establishment of relations of non-domination and creative harmonisation between human and nonhuman nature.

This is what makes post-structuralism potentially valuable. If the governance of disciplinary norms is a permanent feature of our lives, if state and non-state agencies reach into the capillaries of the social body, and vice versa, then ethical discourse must aim not at the elimination of governance but at the greening of governance. Relations of normal deviation and abnormal deviation may be susceptible to collective tectonic shifts away from our existing productivist imperatives towards environmental ones. As such, the 'abnormals' would not be those who fall, or sometimes jump, outside the race to earn/consume/earn/consume, but those who live and work unquestioningly without reference to performative debate and ecological critique. Unlike Wissenburg, then, for whom preferences and norms are given, a discursive Green democracy requires the association of autonomy with its natural pre-conditions.

The second aspect of discursive reflexivity can be dealt with more briefly as it is being covered extensively in the literature on reflexive modernisation and risk society. This debate proposes that the consequential ghosts of the first phase of modernity return to haunt the second phase of modernity (Beck, 1992; Giddens, 1994). The scientific techno-rationalism of the industrial era, based upon the hierarchical production of goods by social experts and their consumption by the mass of non-experts, is replaced by a post-industrial risk society where the generalised experience of hazards and 'bads' makes each person into both an expert and a neophyte in the ways of the world. Most obviously, the attempt to externalise, master and dominate nature has led to a feedback mechanism of ecological damage and catastrophe where nature shows us that we are not so clever after all (Beck, 1995). This means that the traditional goal of progress (the elimination of bads at source) has to be replaced by a strategy of risk management, of living with ambiguity, uncertainty and the permanence of ever-accelerating change. In a risk society, the centres of power and responsibility have vanished, leaving citizens who are more highly empowered than ever in a world that resists control with greater ease than ever before.

This means that policy-making can no longer take place behind closed doors but must be conducted in the open. And unlike the social reforms of the modernistic welfare state, where those who designed policies and institutions were often not those who would have to experience them, a greater dialogue must be established between specialists and non-specialists, with the realisation that the latter are no less experts than the former in that they possess knowledge and experience with is *different* rather than inferior. At local, regional, national and international levels, then, there must be formal and transparent systems of reciprocal interaction between specialists and lay opinion, systems that are nevertheless receptive to the informal protests of oppositional movements that will always confront the comfortable certainties of both sides. Such proposals are already on the agenda to some extent. For instance, the agents of globalisation such as the World Bank and World Trade Organisation are, in the wake of widely publicised protests in 'global cities' around the world, more open to discursive contestation than ever before; similarly, the politics of food production has become open to the public gaze due to various health scares and the failed attempt of corporations and governments to hide developments in genetic modification.

Economic democracy

The literature concerning ecological democracy has made few connections with that concerning economic democracy (cf. Gorz, 1989). This is partly the fault of Green theorists, many of whom often seem split between an ownership-as-usual approach to ecological reform and an anarchistic approach that bypasses the issue of property rights altogether. Many socialists, too, have ignored Green concerns. For instance, the revival of interest in market socialism in the 1980s and 1990s said very little about sustainability issues.

Interesting connections are there to be made, however. At a simple level it can be claimed that socialism needs ecologism if it is not to repeat the error of trying to beat capitalism on its own terms, i.e. in terms of growth, efficiency and productivity. Instead, socialism requires new organising principles that propose qualitatively different goals and values, and Green economics offers an obvious resource in this respect. Conversely, ecologism requires socialism if its promises of mutuality and discursive democracy are not to founder in an economic system distinguished by imbalances of economic power. As Offe (1996) suggests, welfare state expansion stalled because its emphasis upon altruism and collectivism collided with an economic system based upon gain and self-interest – with proposals to nurture socialised forms (as opposed to statist forms) of property coming too late in the day, e.g. the Swedish Meidner plan – and there is little reason to imagine that Green reforms would be any different. Nevertheless, if the socialised ownership and control of natural and social resources is to be placed on the political agenda then social policies may have crucial role to play for two reasons.

John Roemer (1994) has spent several years elaborating the outlines of a market socialist economy. This economy would contain two types of money: commodity money for the purchase of consumables and share money (coupons) that are issued by the state treasury. Coupons could not be sold themselves but they could be used to purchase shares in public firms, shares that generated a 'social dividend' or stream of cash during an individual's adult life. Roemer is therefore proposing a socialisation of a large part of the economy that manifests itself as a dividend for each individual (the value of which will vary): a market socialism that utilises forms of both individual and collective ownership. This idea suggests two points. First, perhaps two types of coupon can be envisaged: *social* coupons that can be used to purchase shares in firms and *natural* coupons that can be used to purchase shares in land. But whereas the value of the social dividend is straightforwardly proportionate to the productivity of the object of investment (the firm), some way would have to be found of making the value of the 'natural dividend' proportionate to the sustainability of the invested land. For instance, it is often alleged that a sustainable economy would lead to improvements in health due to reductions in stress, depression and pollutants; if so, then the value of the natural dividend could be related to such monetary savings via Green indicators such as the Index of Sustainable Economic Welfare. Secondly, Roemer's social dividend can be thought of as a more elaborate version of the Basic Income proposal discussed in Chapter 8. The implication is that appropriate reforms of the income maintenance system might present us with an embryonic version of the social and natural dividends proposed here and so a potential platform from social security to social ownership and control.

The second relevant policy area concerns pensions policy. Even putting ecosocial issues aside there is a compelling need to democratize the occupational and private systems that characterize a growing element of the world economy (Fitzpatrick, 2001a). As people are called upon to provide for their own futures away from social insurance provision, so contributors should be able to own and control those systems in conjunction with professional fund and portfolio managers.[5] At present, such pensions are 'passive', with limited amounts of information available as to their scope and financial 'location'; more 'active' systems must therefore be legislated for, with members having the kind of collective and democratic powers regarding investment decisions that has hitherto been denied. Governments have been reluctant to follow this train of thought as such democratization potentially represents a form of economic socialization. With members in control, investment decisions would either be based upon a deeper ethical dimension than at present or nonethical decisions would be more accountable to public opinion than they currently seem to be by occurring in the democratic light of day. (The closest that UK politicians have come to contemplating radical reform is with reference to the Singapore Central Provident Fund: a potential advance on today's *non-state* schemes but one that treats members largely as consumers and as units of national economic management.) It would then be relatively easy for governments to reward, e.g. through tax breaks, ethical decisions that embody the principle of sustainability. Regulating the market in this way could produce a virtuous circle of incentives and returns, an ecological nonzero-sum game compatible with today's political economy but one that potentially points beyond the limitations of market capitalism: a transformatory stance that defines the rationale of ecosocial welfare. Obviously, this only scratches the surface of an interesting, important but immensely complex idea. Indeed, devising proposals for a Green economic democracy would probably be a full-time job for a research institute. Even so, ecosocial welfare must make reference to these kinds of ideas if it is to offer anything more that a revamped social democratic welfare state.

Conclusion

Can ecosocial welfare resolve the two contradictions of the welfare state? No, it cannot. Its attraction is that it plants one foot, *but only one*, in the realities of the limited present, but this also means that the first contradiction, between social legitimation and economic accumulation, and the second, between growth-driven distributive justice and natural preconditions, can be healed but probably not cured. Because it is concerned with the practical and the do-able, ecosocial welfare does not make any of the grand, world historical claims that have previously been proposed by radicals as means of resolving the inherent contradictions of capitalism. However, ecosocial welfare's other foot stretches away from the unimaginable present into an agathatopian future where legitimation, accumulation, justice, growth and sustainability begin to converge. The theory of ecosocial welfare is both a thought-experiment and a potential site of ideological mobilisation that is designed to test the waters of political imagination within the borders of social policy. Disagree with the details, by all means, but for what reasons can you tell me that the experiment is not worth performing?

5
Local Welfare: State and Society

Michael Cahill[1]

The decade since the Rio Earth Summit of 1992 has seen an explosion of academic work in the social sciences around the concepts of sustainability and sustainable development with a related focus on the meanings of intergenerational equity and environmental justice. The locality featured prominently in the declarations adopted in Agenda 21 at the Rio Summit in 1992, most notably in Local Agenda 21. It has been a long time coming but gradually the sustainable development message of Agenda 21, that environmental and social policy needs to be brought together, has been accepted by pressure groups, NGOs and, since 1997, by the UK government. The subject area of Social Policy and Administration, which conventionally has taken as its focus the work of the welfare state, has begun to incorporate some of this work. The UK welfare state which emerged in the 1940s was a statist version of welfare which displaced and replaced other versions of collective provision. It was, in its own time and on its own terms, extremely successful but its success obliterated the memory of alternatives which had existed previously. This chapter takes as its focus the idea of a local welfare state, considering the historical origins of the idea, examining the socio-cultural climate which determines the range of responses to social problems in the modern locality and concluding with an argument for the importance of the recent attempts to reinsert ideas of welfare society into the debate

The importance of the local

Size has always mattered for Greens and small has always been beautiful; among the numerous sketches of the future Green society there is one common element: it will be a small scale local society.

This is to be found in the work of William Morris (1890) – who is now claimed as the world's first eco-Marxist – notably in his utopian novel *News from Nowhere* published in 1890 but set in 1953 – where the cities and towns have been deserted and the population after the British Revolution live in small rural-based communities. The German Green Rudolf Bahro writing in the 1980s believed that the ideal size for a community should be between 10,000 and 3000 people (Bahro, 1986: 92–8). Bahro's communities, similar to those dreamed of by Morris, would be post-industrial in the sense that they would live off the produce they grew from the land and they would have foresworn many of the products of consumer society. This presumes a value consensus among the local population which it is difficult to envisage being achieved in our diverse and pluralist societies.

Another strand in the Green emphasis on localism is the equation of the local with the natural. Bioregions are defined as areas with natural characteristics of climate, geology and flora and fauna which together form an ecological sub-system. With authors such as Devall (1990: 58–9) the assumption is that people in the bioregion will be sensitive to the needs of that region and so develop an ecological consciousness. This is a large and unrealistic assumption, and as Eckersley (1992: 169) has pointed out 'linguistic, religious and cultural boundaries do not necessarily follow bioregional lines'.

The arguments for the importance of the local in the Green tradition are also to do with the small scale being that much easier for people to relate to and for local people to control and manage. The importance of face-to-face contact is important in the building of community at the neighbourhood level, often involving the creation of organisations, clubs and associations. Equally important is the environmental case for a local economy, where food is grown locally thus obviating the need for long distance transport of foodstuffs, and for local communities to sustain themselves through their own production of goods and services. Nonetheless there is obviously no unchanging essence to the local – it has changed markedly from the time of Morris. Indeed, we can say that the idea of the local as removed from the rest of civilisation is no longer the case. Global media and communications put the world within the reach of most people in rich countries.

The arguments around the importance of the local nowadays have a sharper inflection because of resistance to globalisation. This has led to a call for 'localisation' as an alternative to the mainstream political acceptance of globalisation. This would entail localisation of the economy, of the polity and of the welfare state. Going counter to the assumed consensus that globalisation is irreversible it envisages countries where trade is predominantly local as is employment and food production. In some of its aspects it is a return to the alternative economic strategy of the 1970s, with a call for nation states to be able once more to erect trade barriers and impose import duties which would protect their local industries against free trade (Hines, 2000). The localisation of welfare existed before the First World War and it has been canvassed as an alternative to state welfare in the UK since the early 1980s. Before examining the prospects for local welfare today one needs to consider the historical record.

The demise of local welfare

There was a ferment of thinking around the proper role of the state and the individual in the provision of welfare before the First World War (Gilbert, 1966; McBriar, 1989; Harris, 1994). It would be foolhardy to claim that the versions of the local welfare state current before World War One might be seen as precursors of the ecological approaches to welfare currently on offer. It is, however, useful to cast some light on why they were unsuccessful.

The Fabians Beatrice and Sidney Webb were articulate exponents of municipal welfare in the period before World War One. During their long career they witnessed the Labour Party grow from a small minority presence on a number of town and city councils to forming the national government. They played an important part in the period before World War One in the formulation of socialist thinking about the role of the municipality, especially in relation to social policy with their Minority Report on the Poor Law Commission which recommended that local authorities should assume the functions of the Poor Law. Later Fabian versions of social policy assumed that for services to be delivered on an efficient and equitable basis central government would need to play the commanding position. Hence the nationalisation of health care in the 1940s, the further consolidation of a national system of education from the 1940s onwards, the National Assistance Act, and the centralisation of social security benefits (Loughlin, 1985). Local authorities witnessed the removal of the municipal hospitals which they had been responsible for since the demise of the Poor Law in England and Wales in 1929; the friendly societies saw their role as insurers for working people being assumed by the state; and numerous voluntary organisations found that their social work was declared redundant in the era of the welfare state.

This is not to deny that the creation of the 1940 welfare state was not associated with a clear improvement in the level of health care, the range of social security benefits and the nature of social service support. Rather, it is to note that the centralisation thrust did eclipse a variegated and substantial social policy culture in the locality where it was more amenable to local and democratic influence. Power was transferred not only from local associations and local government to central government but also from volunteers to professionals. This is where the origins of the academic study of Social Policy and Administration lie.

Social administration and the professionalisation of welfare

For the greater part of the twentieth century social administration was only taught to social and voluntary sector workers. In the United Kingdom it emerged from the teaching and practice of social work and the empirical study of poverty. The courses which were established for the teaching of social workers and other welfare workers in the universities at the beginning of the twentieth century were the beginnings of the professionalisation of social and charitable work. From this point onwards it was possible for young women, and a few men, to make a career in social work. It took forty years for the professional to supplant the volunteer in the direction and organisation of social services however. This work was in direct line of descent from the perception of the role of the subject held by Sidney and Beatrice Webb who had founded the London School of Economics in 1896 (Bulmer *et al.*, 1989).

Under the leadership of Richard Titmuss social administration was established as a subject area in its own right and it acted as a handmaiden to the emergent welfare state services in the post-war period. The advent of professionals in the public services added another (usually articulate) participant in the discussion of local policy. Social workers, doctors, nurses, teachers and public health staff were all employed in various guises by local authorities. As the twentieth century progressed then it can be observed that their loyalty was often more to their own professional body and work rather than to the local authority which employed them. This rise of the professional class was mainly seen in a favourable light (Perkin, 1989). It was not until the work of Illich that the disempowering nature of professional activity began to be acknowledged (Illich, 1977).

But what were the varieties of local welfare provision that were lost and what are the alternative visions?

Local welfare in the late Victorian and Edwardian age

The localist versions of voluntary welfare in which philosophies of citizenship emerged in the period up until the First World War was a product of a world where capitalism was locally based. The assumption by politicians was that welfare was best provided locally, as it had been for centuries by the Poor Law, while the new services which were being introduced outside of the Poor Law – school meals, school medical inspection, children's welfare – were to be provided by the local authority.

Looking back one hundred years later it does well to remember that on the verge of the twentieth century there was a consensus across the political spectrum that the future was to be local. William Morris and E. Belfort Bax writing in their socialist 'textbook', *Socialism: its growth and outcome*, believed that the national political system should be 'starved out':

> there should take place a gradual and increasing delegation of the present powers of the central government to municipal and local bodies, until the political nation should be sapped, and give place to the federation of local and industrial organisations. (Morris and Bax, 1893: 282)

Admittedly, Morris and Belfort Bax were on the extreme left but Fabians too saw local government as central to the realisation of socialism. We find Bernard Shaw writing in *Fabian Essays* that

> Local self government remains prominent within the sphere of practical politics. When it is achieved, the democratic state will have the machinery for socialism. (Shaw, 1889: 188)

The Fabian Society published hundreds of 'how to do it' pamphlets on all aspects of municipal socialism. The pressure for municipal powers and enterprise came from Liberals and Conservatives as well, although not to the same extent. The dividing line between reformers and conservatives on this issue was the extent to which the former wanted to extend the concept – towards issues of social policy to school feeding, medical inspection and municipalisation of health services including the local hospitals. This debate went on within a context of civic consciousness in which cities vied with one another for prestigious municipal buildings, parks and other municipal schemes. It was, in marked contrast to our own time, a period in which the locality was the centre of political activity because these were local economies and societies: it was a local capitalism. The First World War changed this, creating a national society and hastening the growth of a national economy and greatly increasing the powers of central government. But municipal social welfare had been very shaky before the First World War as ratepayers were reluctant to finance the relevant costs, leading to widespread calls for central government financial support.

Several conclusions can be drawn. One is that local government needs its own source of revenue – the Layfield Committee on local taxation established in the 1970s recommended a local income tax (Layfield Committee, 1976). Without it there is an inexorable trend towards the centralisation of local services. It does well to bear in mind that this local welfare provision came out of associational culture. As José Harris has noted:

> Throughout Britain in the 1890s and 1900s working people above the level of the very poor were involved in a richly variegated, autonomous and self-governing associational culture, expressed through trade unions, friendly societies, co-ops, chapels, and political and social clubs. (Harris, 1994: 193)

Among the varieties of local welfare schemes, which included various attempts to revitalise philanthropy and to put new life into charity organisation, there was an effort to theorise the necessary changes towards a society which could best respond to the social problem of poverty and its associated effects in poor educational attainment, poor housing and poor health. In the writings of various New Liberals and other committed social reformers one finds the stress on citizenship as a way for the middle class to reach out and befriend those in poverty and for the working man to grasp some of the educational and other opportunities on offer. Before the First World War questions of wealth and poverty could be contextualised within the compass of a single city – citizenship really did then bear a relationship to its original meaning as someone who had the rights and duties of a member of the city. Perhaps the most optimistic conclusion to be drawn from this brief excursion into the historical record of the locality is to say that that the Whig interpretation 'from Victorian city to Welfare State' was not inevitable and to remind ourselves that other countries have decentralised welfare systems.

Given the experience of the twentieth century one can justifiably question whether localising the centre is particularly effective. Bureaucracies are not only a national phenomenon and local bureaucracies can be more difficult to control. Equally the focus on the local is a problematic one, for in this age of personal mobility some people will spend much more time out of the geographical area where they happen to live – not only at work but also with friends and family. Their knowledge and identification with the locality may then be limited and will not provide a resource for them to draw upon. Such observations have been made in relation to the revival of community by the communitarian movement in the US and in Western Europe. As Rodger (2000: 185) has recently observed important debates are now in progress about character, social capital and community, revolving around the concept of welfare. One might add that they have actually *resumed*, because with due alteration of detail such debates were a recurrent feature of the Edwardian and late Victorian polity. We have now moved beyond the caricature of social welfare which was in vogue twenty years ago, in which charity workers were agents of social control and voluntary action was seen as no more than paternalism.

Clearly one needs to keep this history in perspective. The context was one of local capitalism in which the ruling class was visible because at least until the late nineteenth century they lived in the city. It does well to remember that the earnest debates which were engaged in by members of the Independent Labour Party and other socialist societies were only a small minority of the working class. This too was a working class where the female franchise was tiny and even the majority of males were without the vote given the operation of the voting system. Cook has estimated that the municipal franchise in this period rarely exceeded 20 per cent of the adult population (Cook, 1976: 39).

The debates about the proper sphere of the individual and the state, many of which in the Edwardian period centred on welfare, were suffused with ethical debate. Ethical questions suffused the work of socialists and reformist Liberals, most notably with the New Liberals. Contemporary discussions in communitarianism and Green theory of the good life and the virtues of citizenship echo those of a century ago. What must also be borne in mind, however, is that there were louder and more effective voices which spoke to the tenor of the times – commercialisation and the beginnings of mass entertainment drowned out the ethical discussion. (For an acute analysis of this process in Reading in this period see Yeo, 1976.) To reassert the importance of the local in a globalising world is a complex endeavour and our system of transport is one reason why.

Mobilities

The motor car which, in Edwardian England, was the preserve of the rich and driven for the most part by their chauffeurs has become the transport of choice for the bulk of the adult population a century later. Private transport is now the constitutive basis of our world and it has been a major agent of delocalisation. The dominant world order is built on the premise of mobility, with information technology and transport providing its foundations – it is the free movement of goods which is an essential plank in free market ideology. They provide the scapes – the networks – which underpin financial markets, global media and increasingly the production of goods and services (Urry, 2000). The car has transformed cities, displacing large numbers of people in urban areas, increased travel distances to work and to leisure facilities and made possible all sorts of previously impossible journeys, thus opening up the countryside to urban lifestyles; it is also a major contributor to carbon emissions which threaten to severely disrupt the world's economies and societies (Whitelegg, 1997). The technology of the car has radically transformed the ability of individuals to access places and people at times of their own choosing – it is inextricably connected with modern ideas of freedom (Urry, 2000: 59–60).

In a consumer society access and participation are all important. To be a consumer-citizen one has to be a car driver and those who lack this ability are second class citizens denied access to places where their own legs, whether aided by public transport or the bike, will not take them. This fact has been understood by many young people – a recent research study on young people's attitudes reported that the driving licence was regarded as much more important than the ability to vote (Solomon, 1998). Full citizenship and participation in the consumer society is built around mobility, including the ability to access jobs, many of which are now in decentralised locations. As the evidence on the link between lack of cars and social exclusion mounts then governments respond by stressing their desire to see more and more people learn to drive – a clear example of environmental priorities being subordinated to the hegemony of the car. Less private cars in poor neighbourhoods means that one approach is to promote car purchase in socially excluded areas. This 'helps' the individual but damages the environment and it will further exacerbate the lot of the other non-drivers who will see a further deterioration in their bus and train services.

A more democratic and local approach to transport would be to prioritise the environmentally benign modes of cycling and walking. If these are made safe and secure from motor transport then large numbers of journeys, which at the moment are not made because of the fear of traffic, can be contemplated. The large numbers of children who own bikes but who do not use them as a form of transport are a prime example.

Consumer citizenship

As the dominant ideology of our age, consumerism inflects the discussion of citizenship as with so much else. This was seen with the Citizens Charter of the Major government where the last remedy for disgruntled citizens dissatisfied with the performance of a public service was money. Generations who have been brought up on ideas of themselves as consumers do not readily appreciate ideas of the 'public interest' or 'academic community' preferring instead to view their contact with the public sector as a consumer/seller relationship. As has been recognised this has affected views of welfare state provision. Yet there are also more wide ranging consequences which flow from the dominant metaphor of the market. Citizenship, for example, has been largely seen in terms of rights and it is this which provides the concept with a resonance in a consumer society.

Consumer-citizenship is not about conventional political participation in the form of political party membership, voting at elections or political news. Consumer society promotes the realm of the private individual world as the reality which one can alter and enjoy so that the public sphere is necessarily regarded as alien and unimportant.

A necessary caution here is that consumer society has the ability to incorporate radical ideas. Arguably this has occurred with some versions of feminism and alternative life styles (Klein, 2000). Those who advocate Green citizenship are going against the grain of our times, they can be said to be saying that we should have less fun, and they can be portrayed as questioning the empowerment afforded by consumer goods and consumerist rights: the freedom to drive, the freedom to travel, the freedom to do what they like within the law. The promise of freedom for most people is the consumer market place. Consumption is the glue which keeps individuals as part of social groups and acts as a bond in peer group activity; consumption is a form of signalling one's likes and dislikes to the wider world and of winning acceptance. But as in the example of children being able to ride their bikes in safety, and thus seeing friends and exploring the area where they live, there are many forms of satisfaction and enjoyment offered by a more sustainable lifestyle. We are bound to consider whether consumerism has numbed the awareness of people to environmental crisis. This means exploring the reasons why our relationships with each other and the planet may be in need of repair.

Post-emotionalism

Why is it when each week brings further news of the acceleration of climate change, environmental disasters, species extinction and innumerable other instances of damage to the ecosphere, that there is so little public protest and anger? Post-emotionalism furnishes us with one part of the answer for it points to the reasons why our society and culture appears to be in denial about the environmental crisis. Mestrovic (1997: xi) argues that emotions in consumer society are manipulated by marketing and media working for dominant economic interests:

> synthetic, quasi-emotions become the basis for widespread manipulation by self, others, and the culture industry as a whole.

This permits consumer capitalism to invade parts of the individual psyche in previously undreamt of ways. The neglect of serious and pressing environmental issues by the great majority of the population in rich societies must also be the result of the way in which the mass media are used to disseminate information. Neil Postman in his book *Amusing Ourselves to Death* predicted that public information and news would be trivialised because of the way in which it was handled by broadcasters (Postman, 1987). This is not only a question of selling more goods but it is a cultural change which has entered deep into the psyche of the majority of people in our privatised world where – deprived of the meaning afforded by religions or political ideologies – they make sense of their lives via the symbolic interpretations offered by consumer capitalism.

Mestrovic has identified a central issue in the manipulation of people in contemporary society with his ideas on post-emotionalism:

the ways in which identity and self are targeted by advertising, public relations agencies and management techniques which encourage ideas of participation in the workforce. The argument is that emotion has been turned into a 'quasi-intellectual phenomenon' that can be manipulated (Mestrovic, 1997: 38). The appeal of post-emotionalism is not confined to large corporations or their allies in advertising but is used across the political spectrum. A post-emotional society is one where emotions are the very stuff of soap operas and daytime television and where the response to almost every problem is therapy and counselling. But what is important for this discussion is his claim that post-emotional people – and the majority of those in consumer societies fall into this category – are capable of feeling a great variety of emotions but they are disconnected from action. This is because they have reached the conclusion that they are powerless to change politics or world events. When the world's governments have only produced a weakened version of the Kyoto agreement on reduction in carbon emissions, and when the fate of the planet never made the agenda in the 2001 UK General Election, these ideas have obvious resonance. Individual behaviour is crucial to the operation of local welfare and to the success of welfare associations and other local forms of welfare. The same broad trends which Putnam and others have identified in the United States are at work in the UK (time deprivation, withdrawal from voluntary organisations, declining support for Parent Teacher Associations) and means have to be found to reverse these trends.

Moralities

The operation of welfare requires a body of opinion which supports welfare values, just as environmentally responsible behaviour requires that underpinning. Yet we live in a society where meaning and image are manufactured to boost corporate and commercial aims. The structural determinism which has been so pervasive on the left in UK politics has minimised the role of human agency, while corporate capitalism has been content to ignore the accumulating environmental problems. But the severity and increasing frequency of environmental crises impels us to consider individual behaviour. As Redclift has observed:

> the human behaviour which underlies global warming is rarely considered. More attention is paid to ways of mitigating the effects of global warming, than to its causes in human behaviour and choices,

the underlying social commitments which make up our daily lives. (Redclift, 1996:17)

The ecological perspective comes out of a long tradition stretching back to the eighteenth century which presumes some knowledge of and facility in reasoning around the conceptual basis of freedom, community, democracy. There has been a remarkable body of work which has delineated the connections between humankind and the natural world exploring the extent to which human beings can relate to other parts of the animal world and take their interests into account in decisions about nature (Barry, 1999a). The transition to a sustainable society will require society to put together the pieces of moral discourse and use them as a basis for new understandings.

Personal responsibility

The instrumentalism to be found in personal relationships is a barrier to personal responsibility. It leads to individuals and families putting their own wants and needs before the common good. A society which wanted to move seriously to a less environmentally damaging condition would need more, not less, rules. Individuals would need to take their own personal responsibility towards the environment seriously: in relation to recycling, using the car less, producing less household waste. At some point there will need to be restrictions on car use if rich societies are to comply with internationally agreed carbon emission targets and it is much better if these are regarded as desirable by the population. Here then lies the importance of the state and the public realm for the promotion of such values but this has to be done in a positive manner and not with the state acting solely as the environmental police.

In a sense this is what the popular communitarianism of Etzioni was about: trying to work out for different communities of race, age and gender their common moral agreement on certain values (Etzioni, 1995). This is a necessary endeavour for government and for all those concerned with the provision of services to their fellow citizens, as the transition to a sustainable society cannot be an imposed process but must be based upon the greatest possible consensus. If opportunities can be found for debate of this kind then some form of critical distance can be opened up between consumerism and everyday life.

A strengthening of environmental responsibility would need to build on some of the discussion around communitarianism. The vision of consumerism built on freedom of choice, an unending vista of material progress, gobbles up resources and also gives an air of unreality. A sustainable society needs to provide its members with emotional attachments and a stable base (see Kraemer, 1997). The security and sense of worth and value which come from loving relationships are easily lost in a world where freedom of choice as an ideal is transferred from the economic marketplace to the world of personal relationships. Family structures have weakened in the climate of personal freedom which, in part, has been fostered by consumer capitalism. Choice, which is such a totem in the market place, has become a key element in personal lives so that many men and women who are parents believe they have a right to break up a marriage if they feel they are not getting the fulfilment they require. The high divorce rate since the early 1970s has produced in succeeding generations a fragility about relationships which means that far fewer people are now prepared to commit themselves to marriage. Consumer capitalism has also led to a position where paid work has assumed quite considerable importance in people's lives, thus reducing the opportunities for voluntary work and civic engagement. The model of ecological stewardship (Barry, 1999a: 255–61) premised on the cultivation of virtue requires a mature personality which integrates and resolves – as far as any one can – their impulses and drives. Consumerism with its implied promise that all will be well if a particular product or look is purchased does not aid this process.

It would seem that there is a process of civic disengagement at work in our society in much the same way as that identified by Putnam in the USA (Putnam, 2000). The 'work and spend' society has had a pervasive influence on social attitudes which has affected caring, this being at the heart of welfare. Those who work in front-line social and health care services – day care, domiciliary help, nursing – have seen their status and remuneration decline so that these areas of employment are now unfashionable and seriously short of experienced staff.

Sustainable welfare

There is a natural affinity between social policy and ecological modernisation (see introductory chapter). Both are concerned with the pragmatic and the do-able. This is why Fabianism was for so long the unofficial ideology of social policy: it proposed concrete solutions, its proposals were rooted in the politics of the day, it was ameliorative and it would improve the lot of ordinary people in the here and now: 'pragmatic, policy-oriented and anti-ideological' (Bluhdorn, 2000: 191) is a description of ecological modernisation but it fits social policy equally well. But this perspective does not create the values and culture which underpin reciprocity, caring and neighbourliness. Rodger (2000) has argued for the need to create a principle of social cohesion which will permit the creation of a welfare society.

There needs to be a substantial and articulate body of opinion supporting these environmental and welfare measures otherwise there is a real risk that Green thinking will be an ecological gloss upon a society whose structures and processes are fundamentally environmentally damaging. An environmental ethics would underpin Green citizenship, and the creation of such a body of thought will, in part, be produced by the mass media, given the nature of political communication in our society, but it will also require self expression and this will need to come from the locality. It is here, then, that associations will play their part for it is through the creation of voluntary organisations and associations that abstract concepts such as sustainability create their meaning for most people.

The politics of Agenda 21 embody the international dimension of the Rio sustainability summit and the local aspirations of communities as embodied in local government policy making. Although it has been implemented in most local authorities the amount of resources allocated to the programme has been small scale and inadequate for the ambitious targets set. In its new guise as part of the Community Strategy it may well assume a greater importance in the strategies of local authorities. It has performed a number of valuable roles in the promotion of sustainable development: popularising the social economy of LETS (see Chapter 9) and credit unions, linking discussions of main local authority programmes such as transport and housing to sustainability and furthering environmental education. All over the country there has been discussion of the role and significance of sustainability indicators and a debate about the dimensions of 'quality of life'. This 'social sustainability' dimension is a visible way of broadening the concerns of social policy to embrace the environmental agenda. There is widespread agreement that local government needs to be modernised, that its procedures are outdated and derive from nineteenth century models of the local democratic process.

Clearly, there needs to be change if low turn-outs at local elections invalidate the basis of elected representatives but we must be careful that the changes do not merely import a range of management techniques which might make local authorities more efficient but much less democratic. The 'crisis of the welfare state' which emerged in the late 1970s was not only about the ability of state expenditure to finance welfare needs, it was also a crisis of legitimacy. The loss of public support for some welfare state institutions led to a rethinking of appropriate structures for welfare delivery. One of the manifestations of this in local government was the revival of ideas of local or municipal socialism in a number of Labour authorities (Gyford, 1985; Stoker, 1987, 2000). These and subsequent decentralisation initiatives foundered on the realities of a local government system being progressively controlled from the centre in the pursuit of public expenditure reductions. Decentralisation as a way of organising local authority services has not, however, disappeared from the agenda of local government.

One of the considerable merits of Paul Hirst's proposal (that self-governing voluntary associations, receiving public funds proportional to the number of their members, should provide welfare services) is that they would reflect the diversity of cultural, social, religious and political groups today. Such a scheme would also enable a degree of choice for members and the ability to exit an association if a member was dissatisfied. As Hirst points out there would still need to be a role for the state in setting standards of service, the quality of service and the level of entitlements (Hirst, 1994). Associations might also begin to tackle some of the most pressing problems affecting low income areas such as the lack of shops selling fruit, vegetables and groceries. The 'social economy' of LETS, credit unions and time-banks has the potential to link with associational welfare. Indeed, there are many historic parallels between this work and that of the Co-operative Societies.

There needs to be a reconnection with social philosophy, involving an examination of the nature of the good society and the good life which involves some analysis of the cultivation of the virtues. Green citizenship makes sense where the virtues of caring and altruism can be practised at the local level. This provides a point of connection with the pioneers of social administration, such as Tawney, who believed in virtuous people and good citizens.

6
Quality of Life, Sustainability and Economic Growth

Tim Jackson

The last hundred years have seen a massive increase in the wealth of this country and the well-being of its people. But focusing solely on economic growth risks ignoring the impact – both good and bad – on people and on the environment ... in the past, governments have seemed to forget this. Success has been measured by economic growth – GDP – alone. We have failed to see how our economy, our environment and our society are all one. And that delivering the best quality of life for us all means more than concentrating solely on economic growth.

Tony Blair, 1999

Introduction

Tony Blair's introduction to the UK Government's sustainable development strategy *A Better Quality of Life* (Department for the Environment, Transport and Regions, DETR, 1999a) represents a startling departure from conventional thinking about the complex relationship between national well-being and economic growth. It is a departure, moreover, whose full ramifications – familiar though they may be to the Green lobby – the Prime Minister himself may not yet entirely have taken on board.

A Better Quality of Life provided the basis for a second document called *Quality of Life Counts* (DETR, 1999b) in which the Government established a set of 15 headline indicators of sustainable development. At the top of this list, of course, is the familiar indicator of economic growth, closely followed by indicators relating to investment and employment – also key concerns of traditional government. But the list includes, in addition, indicators related to educational qualifications, social conditions, environmental impacts, resource use and even certain aspects of biodiversity. The intention of the sustainable development strategy is that these indicators should make up a 'quality of life barometer' which will be used to measure 'overall progress', a way of ensuring that we are achieving a better quality of life for everyone, now and for generations to come.

But what exactly is quality of life? How should we ensure that it is maintained or improved? And why does Tony Blair's introductory comment herald such a radical departure from the received wisdom of previous governments? The answer to this last question lies in part in identifying relatively straightforward answers to the first two questions, answers which have dominated conventional economic and social policy during most of the last century.

Specifically, successive governments in most industrialised nations have for decades taken quality of life to mean more or less the same thing as standard of living. The standard of living in its turn has traditionally been equated with per capita levels of national income; and national income is measured conventionally through GDP. Since GDP rose more or less consistently over the last 50 years, the comforting logic of the orthodox view suggests that we have been pretty successful in delivering an increasing standard of living and, by proxy, an improving quality of life in the UK over recent decades. Furthermore, if our concern is to ensure that quality of life continues to reach new heights, the conventional view provides a ready and familiar formula for achieving this end: namely, we need to ensure high and stable levels of economic growth.

Of course, it has sometimes been recognised that average per capita income may hide a significant individual discrepancy: some earn much more and some earn much less than the national average. But then the correction of such discrepancy, or at least the amelioration of its most pernicious effects, is precisely what social policy has always been about. By earmarking a proportion of the national income for public spending, successive governments have attempted (sometimes successfully) to maintain law and order, provide state-funded education, build-up, re-structure (or occasionally just run down) the health service, and offer minimum levels of social security to those who, for one reason or another, find themselves unable to earn a decent living.

Thus the conventional picture is one in which social policy (cashed out mainly in terms of a progressive redistribution of incomes) and economic policy (cashed out mainly in terms of the pursuit of economic growth) have trundled hand-in-hand along the highway to utopia: a land where individual quality of life goes on increasingly endlessly for all its lucky inhabitants.

Tony Blair's comment clearly signals a recognition by government that there may be flaws in this happy wisdom. It is a recognition that is long overdue. The truth is that the popular equation of economic output with quality of life has come under increasing scrutiny over the last few decades, from a number of different quarters and for a variety of different reasons. In fact, it would be fair to say that it has never been entirely accepted. The economist Simon Kuznets (1971), one of the key proponents of the system of national accounts, came to recognise that the welfare of a nation cannot be inferred from a measurement of the national income; Meadows *et al.* (1972) attempted to highlight certain environmental *Limits to Growth*; Scitovsky (1976) described the outcome of the mainstream conception as *The Joyless Economy*; and Hirsch (1977) pointed towards the *Social Limits to Growth*. It is even possible to find criticism of the growth project in the 19th century writings of John Stuart Mill – one of the principal architects of classical economics.[1]

In this chapter, I intend to elaborate some of the key elements in recent Green critiques of economic growth. In fact, in the following sections, I delineate several progressively radical attacks on the conventional equation of GDP with quality of life. The first of these resides more or less entirely within the boundaries of neoclassical economics; the second extends the boundaries of conventional accounting to include environmental and social costs (and benefits); the third challenges more fundamental aspects of the orthodox rationale. In the final section of the chapter, I attempt to elucidate some of the challenges those critiques present for modern social (and economic) policy

Hicksian income and the Net Domestic Product (NDP)

GDP may be viewed (and is conventionally calculated) in three different, but formally equivalent ways. It may be seen, first, as the total of all *incomes* (wages and profits) earned from the production of domestically owned goods and services. Next, it may be regarded as the total of all *expenditures* made either in consuming the finished goods and services. Finally it can be viewed as the sum of the *value added* by all the activities which produce economic goods and services.

Of these three interpretations, it is the second which provides the strongest foundation for a welfare-based interpretation of GDP.

Specifically, the expenditure formulation sums all private and public consumption expenditures and adjusts these to account in addition for the formation of capital, i.e. gross investment, and the balance of trade. The sum of consumption expenditures is equivalent (under certain conditions) to the value placed by consumers on the goods they consume and hence, according to the conventional argument, GDP can be taken as some kind of proxy for the well-being derived from consumption activities.

In formal economic terms, the equivalence of consumption expenditures with consumer values is valid only in perfect, equilibrium markets, and it is well enough known that in practice, markets are not perfect. Moreover, it is clear that public expenditure does not take place in equilibrating markets at all; government spending is not allocated according to market forces but according to the political and social priorities of the day. Throughout much of the latter part of the 20th century, however, the response advocated by economic and social theorists – and in particular by right-wing economic and social theorists – to these market 'failures' was to strive for fewer market distortions: reduced taxation, lower public expenditure, less government intervention; in short to pursue hands-off, *laissez-faire* government. Since this strategy also has the consequence of placing more disposable income in the pockets of the electorate, it has had a strong appeal across the political spectrum.

But the welfare-theoretic interpretation of GDP falls heavily at a number of hurdles other than those associated with simple market failure. Even conventional economic theory recognises that it is not sufficient to attend only to current levels of consumption. Well-being, it is understood, consists at least in part in feeling secure about the future. Thus, future consumption possibilities must also play some part in current well-being. This point was raised long ago by the economist John Hicks who pointed out that the purpose of income calculations is to give people an indication of the amount which they can consume in the present without impoverishing themselves in the future. Thus, true income should be calculated as the amount that a community can consume over some time period and still be as well off at the end of the period as at the beginning (Hicks, 1939). Under one interpretation, being as well off at the end of the period depends *inter alia* on having the same consumption possibilities in the following period. Since these consumption possibilities flow from income streams which are generated by capital investment, this requirement has generally been translated into a demand to maintain capital intact. True income is thus the income in the period less the net depreciation of capital during the period.

At the national level, this suggestion leads us to compute the NDP by subtracting the depreciation of capital assets from the GDP. Hicks' argument suggests that the NDP provides a truer representation of national well-being than does the GDP. In fact, in a seminal paper in welfare economics, Weitzmann (1976) showed that NDP can be regarded as a proxy for national welfare in the sense that (under certain conditions at least) it is proportional to the present discounted value of all future consumption. In particular, therefore, a non-declining NDP can be taken as an indication of non-declining consumption possibilities into the future. Conversely, of course, the pursuit of NDP growth assumes (under this interpretation) a welfare-theoretic justification. Though GDP may be flawed as a measure of societal well-being, an appropriate correction for capital depreciation is, according to conventional economic arguments, sufficient to correct the deficiencies.

Measuring sustainable economic welfare

It is clear that a correction of the kind outlined in the previous section represents only a marginal adjustment to the conventional picture. In 2000, for instance, NDP in the UK, as conventionally calculated, would have differed from GDP by less than 5 per cent. The orthodox view, in which increasing quality of life is correlated with economic growth, might be regarded as surviving this kind of adjustment more or less intact. Equally however, it is clear that such an adjustment fails to exhaust the criticisms which have been levelled against GDP as an indicator of national well-being.

In fact, the literature identifies a wide range of problems with the traditional accounting framework, amongst which the failure to account for the depletion of conventional capital is only one. In particular, as we have already mentioned, average per capita rates of GDP disguise the fact that income is unequally distributed, and there are strong grounds for accepting that inequality in the distribution of incomes has negative impacts on the general level of social well-being (Dalton, 1920; Atkinson, 1983; Stymne and Jackson, 2000). In addition, the orthodox measure fails to account for the positive contribution of certain non-monetarised aspects of the economy such as domestic labour and leisure time; it omits any measure of the depletion of natural resources or any account of environmental damage (such as air pollution, water pollution, climate change or ozone depletion); it fails to account for the welfare impacts of social degradation (such as divorce, crime, unemployment and social exclusion).

At the same time the conventional measure includes a range of expenditures which critics of GDP have deemed to be purely defensive in nature (Daly and Cobb, 1989). An increasing proportion of the national income may be spent on cleaning up environmental damage resulting from the production of goods and services, or on treating illnesses arising from impaired environmental quality or social degradation. These kinds of expenditure may be necessary to defend our quality of life against the adverse welfare impacts of other expenditures. But it is then inappropriate to count both sets of expenditures as positive contributions to welfare.

Unfortunately, this is exactly how the GDP operates. The point was made long ago by former US Attorney General Robert F. Kennedy, who in a speech made shortly before his assassination in 1968, famously remarked that GDP

> includes air pollution and advertising for cigarettes and ambulances to clear our streets of carnage. It counts special locks for our doors, and jails for the people who break them. [It] includes the destruction of the redwoods and the death of Lake Superior. It grows with the production of napalm and missiles with nuclear warheads and armored cars for the police to fight the riots in our city. It counts Whitman's rifles and Speck's Knifes and the television programs which glorify violence in order to sell toys to our children.[2]

In the light of such failures, economists and ecologists alike have been tempted to ask whether or not it might be possible to adjust the conventional measure of GDP to correct some of these shortcomings. In particular an adjusted measure would attempt to account for some of the important social and environmental factors which clearly influence well-being, but which are omitted from the national accounts. According to its proponents, the resulting measure – often referred to colloquially as a 'Green GDP' – might provide a better indicator of the nation's quality of life. At the very least, by making such adjustments it might be possible to determine whether or not a non-declining GDP (or NDP) could after all be regarded as a robust indicator of non-declining social well-being.

Amongst the earliest attempts to address this question was a landmark paper published in 1972 by Nordhaus and Tobin (1972) entitled 'Is Growth Obsolete?'. In that paper, the authors constructed a Measure of Economic Welfare (MEW) by adjusting GDP to account for certain economic, social and environmental factors not normally included in GDP.[3] The results of that exercise indicated that between 1929 and 1965, social well-being – as measured by the Nordhaus and Tobin index – increased consistently, but that the growth rate in MEW was somewhat slower than the rate of growth in GDP. The authors concluded from this analysis that growth was not obsolete; that, on the contrary, it continued to deliver increasing levels of welfare; and that, as an indicator of well-being, GDP could still be regarded as robust.

When Nordhaus (1992) examined the same question from an environmental perspective 20 years later, in a paper entitled 'Is Growth Sustainable?', he discovered that his (revised) MEW began to diverge more substantially from GDP in the later years of the study. Nordhaus attributed this increased divergence to conventional sources like declining productivity growth and dwindling savings rather to the unsustainable use of natural resources. But the importance of the study was already clear enough: by making certain economic, social and environmental adjustments to the conventional measure, it had been possible to show that GDP could not be regarded as a robust indicator even of economic welfare, *let alone* of social well-being or quality of life.

This conclusion has been reinforced by the broadest set of studies to date to address this question. The Index of Sustainable Economic Welfare (ISEW) was first developed for the United States by Daly and Cobb (1990) and subsequently revised by Cobb and Cobb (1994). The starting point for the ISEW is the level of personal consumer expenditure. This basis is then adjusted to account for inequality in the distribution of incomes. Positive contributions to the index are made by adding in the value of domestic labour in the economy, those public expenditures which are deemed non-defensive in nature, and net capital investments. Negative adjustments are made by excluding certain defensive private expenditures from consumption, and by subtracting the costs of environmental degradation, long-term environmental damage and the depletion of natural resources. Table 6.1 presents a summary of the composition of the ISEW with the main rationale for each of the adjustments made.

The results of applying this methodology to the United States revealed a trend in sustainable economic welfare which differed

Table 6.1   Summary of the ISEW methodology

Column*	Item	Adjustment	Rationale
B	Consumer expenditure		Basis for the index
C	Income inequality	Atkinson Index	Accounting for the welfare impacts of unequal income distribution
D	Adjusted consumer expenditure	B*(1-C)	New basis for the index, consumption adjusted for inequality
E	Services from domestic labour	+ ve	Incorporating non-monetarised aspects of the economy
H	Public expenditure on health and education	+ ve	Adding in non-defensive public expenditures
I–F	Difference between expenditure on durables and service flow	–ve	Adjusting for the service value of consumer expenditure
J	Private expenditure on health and education	–ve	Subtracting defensive expenditures
K	Costs of commuting	–ve	Subtracting defensive expenditures
L	Costs of personal pollution control	–ve	Subtracting defensive expenditures
M	Costs of car accidents	–ve	Subtracting defensive expenditures
N	Costs of water pollution	–ve	Subtracting environmental damage costs
O	Costs of air pollution	–ve	Subtracting environmental damage costs
P	Costs of noise pollution	–ve	Subtracting environmental damage costs
Q	Costs of loss of natural habitats	–ve	Accounting for losses of natural capital
R	Costs of loss of farmland	–ve	Accounting for losses of natural capital
S	Depletion of natural resources	–ve	Accounting for losses of natural capital
T	Costs of climate change	–ve	Accounting for long-term (future) environmental damage
U	Costs of ozone depletion	–ve	Accounting for long-term (future) environmental damage
V	Net capital growth	+ ve (–ve)	Accounting for changes in human-made capital
W	Change in net international position	+ ve (–ve)	Accounting for international stability

Source: Jackson *et al*.(1997).
* *Note*: The letters refer to column lettering in the original ISEW (Daly and Cobb, 1990; Cobb and Cobb, 1994).

markedly from the trend in GDP over the period examined (1950–88). While GDP in the United States increased substantially over the period, the ISEW began to level out and even decline slightly from about the mid-1970s onwards. Substantially the same methodology (with one or two revisions and additions) has subsequently been applied to a number of other countries including: Austria (Stockhammer *et al.*, 1998), Australia (Hamilton, 1999), Chile (Castaneda, 1999), Germany (Diefenbacher, 1994), Italy (Guenno and Tiezzi, 1996), the Netherlands (Oegema and Rosenberg, 1995), Sweden (Jackson and Stymne, 1996) and the UK (Jackson and Marks, 1994; Jackson *et al.*, 1997).[4] Figures 6.1–6.6 illustrate the results of some of these studies.

In spite of some differences, there are also marked similarities between these country studies. In particular, each of them suggests that sustainable economic welfare grew more or less in line with GDP until about the mid-1970s or early 1980s; but that the adjusted measure stabilises or declines in the later years of the study, in spite of continued growth in GDP. The reasons for this divergence are complex and differ slightly from country to country. Among the principal factors in the UK, for example, are an increasing inequality in the distribution of incomes over the later years of the study, and the steady accumulation of 'ecological debts' from resource depletion and long-term environmental damage (Jackson *et al.*, 1997). Nonetheless, these studies all appear to suggest that economic growth, at least as presently conceived, is far from able to ensure non-declining levels of sustainable welfare.

These results have been cited by Chilean economist Max Neef as evidence for a kind of 'threshold hypothesis' about the relationship between economic growth and quality of life. Max Neef argues that economic growth may indeed lead to increased human welfare in the early stages of development; but that, beyond a certain threshold, the environmental and social costs of growth begin to overwhelm the positive benefits of continued economic growth. For every society, 'there seems to be a period in which economic growth (as conventionally measured) brings about an improvement in the quality of life, but only up to a point – the threshold point – beyond which, if there is more economic growth, quality of life begins to decline again' (Max Neef, 1995: 117).

If this hypothesis is even partially correct, then it clearly poses some important challenges for conventional economic and social policy. In particular, it goes directly against the received wisdom that economic growth inevitably leads to improved well-being, and raises serious

Figure 6.1   ISEW per capita versus GDP in the USA

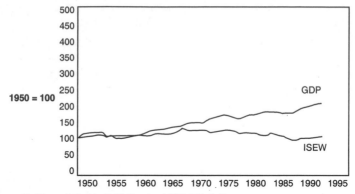

Source: Cobb and Cobb (1994) – in 1972 US dollars.

Figure 6.2   ISEW per capita versus GDP in Sweden

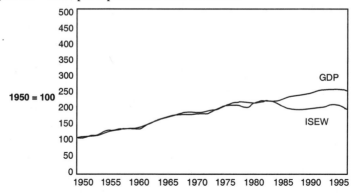

Source: Jackson and Stymne (1996) – in 1985 SeK.

Figure 6.3   ISEW per capita versus GDP in Germany

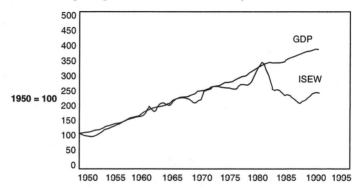

Source: Diefenbacher (1994) – in DM.

Figure 6.4   ISEW per capita versus GDP in Austria

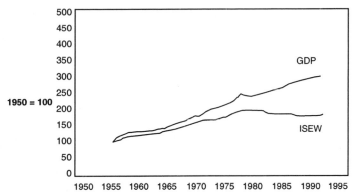

Source: Stockhammer *et al*. (1997) – in 1972 Austrian Schillings.

Figure 6.5: ISEW per capita versus GDP in the Netherlands.

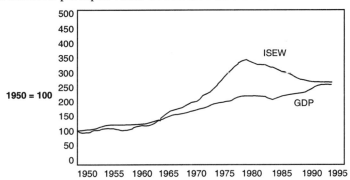

Source: Oegema, T. and Rosenberg, D. (1995).

Figure 6.6   ISEW per capita versus GDP in the UK

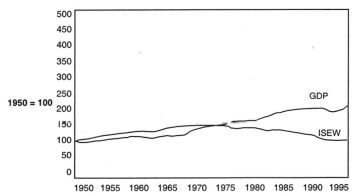

Source: Jackson *et al*. (1997) – in 1990 pounds sterling.

doubts about the assumption that the best way of improving and maintaining quality of life is to pursue policies that will raise the nation's GDP.

Consumption and the quality of life

To some extent at least, the Prime Minister's opening remarks in the sustainable development strategy suggest that this particular lesson from the ISEW – namely its authority as a critique of GDP – has at last struck home within the UK Government.

It is clear that Tony Blair cannot claim precedence for these realisations, even among high-ranking politicians. Robert Kennedy's assault on the primacy of GDP, over 30 years ago, proves the point. The GDP fails to 'allow for the health of our families, the quality of their education, or the joy of their play' insisted Kennedy:

> It is indifferent to the decency of our factories and the safety of streets alike. It does not include the beauty of our poetry or the strength of our marriages, the intelligence of our public debate or the integrity of our public officials. The [GDP] measures neither our wit nor our courage, neither our wisdom nor our learning, neither our compassion nor our devotion to our country. It measures everything, in short, except that which makes life worthwhile.[5]

Perhaps more pertinently, there is a sense in which Blair's remarks fail to match the profundity of Kennedy's critique, a sense in which the government's Quality of Life barometer sidesteps one of the most crucial issues, a sense in which even the ISEW continues to embody the conventional wisdom. For the statistical foundation for the ISEW is still the level of consumer expenditure, disposable income is still regarded as the basis for well-being, and, in spite of Blair's comments, the same assumption still rests at the heart of government policy. The political tenacity of the conventional view is no more clearly illustrated than by the 2001 general election battle between the two main parties. Health and education were certainly raised as issues in the campaign. But the principal contest was still largely fought over rival claims to provide lower taxes and higher disposable incomes.

Christie and Warpole (2001: 63) argue that in spite of its sustainable development strategy, Labour remains politically trapped in the conventional wisdom:

The insight expressed in the Quality of Life strategy, and which is embedded in Labour's best ideas for public renewal, is not yet the organising principle of its politics. Instead it is a proposition undermined by concessions to 'business as usual'.

And business as usual means putting growth first, reducing taxation, further undermining spending on public services and promising yet more cash for the 'voter-consumer' to spend. Increasing the level of consumption remains the single most important political and social goal. Though tempered perhaps by the injunction to account for the environmental and social costs of consumption, Blair's position – and indeed the position of the ISEW – is to maintain the centrality of consumption as the building block for quality of life.

It is precisely this position which is challenged by the more radical Green critique of modern consumer society. In fact, it would be fair to say that critiques of consumption are not confined to the Green lobby; nor are they entirely modern. Even in the nineteenth Century, Marx had already commented on the 'fetishism of commodities'. On the cusp of the twentieth century, Thorsten Veblen (1899) had identified the tendency towards 'conspicuous consumption'. In *The Theory of the Leisure Class*, he contrasted the 'destructive traits' of the 'pecuniary culture' with the 'industrial virtues' of earlier times. In the space of less than a century pecuniary culture had established an iron grip on modern social mores. But it had also generated a host of critics, all of whom were sceptical – in slightly different ways – of the power of increased consumption to deliver ever higher levels of satisfaction.

Lewis Herber (1963: 187) – a pseudonym used by Murray Bookchin in the 1960s – argued that human society had 'reached a level of anonymity, social atomisation and spiritual isolation ... virtually unprecedented in human history'. In attempting to discover why unprecedented and fast-moving prosperity had had left its beneficiaries unsatisfied, Scitovsky (1976) highlighted the addictive nature of consumer behaviour, and its failure to mirror the complexity of human motivation and experience. Fromm (1976) was alarmed at the alienation and passivity which pervaded modern life and placed the blame squarely on an economic system predicated on increasing levels of consumption.

Picking up on Veblen's ideas, Hirsch characterised significant proportions of consumption as 'positional consumption', pointing out that this kind of spending implied certain 'social limits to growth'. As Hirsch (1977: 49) described the problem:

> it is a case of everyone in the crowd standing on tiptoe and none getting a better view. Yet at the start of the process some individuals gain a better view by standing on tiptoe, and others are forced to follow if they are to keep their position. If all do follow ... everyone expends more resources and ends up with the same position.

The obsessive behaviour of the modern consumer, like most psychopathological addictions, fails to generate increasing returns in terms of satisfaction. It simply means running ever harder and faster in order to stay in the same place. As Elgin (1993: 149) argued , 'when we equate our identity with that which we consume ... we become consumed by our possessions'.

These critics have to some extent been supported by empirical evidence. In *The Joyless Economy*, Scitovsky could already cite the failure of reported levels of well-being to match the growth in GDP (Scitovsky 1976). In 1991, Erik Jacobs and Robert Worcester found that people were marginally less happy than they had been in 1981 in spite of increased personal income (Worcester, 1998). A similar result was reported over a longer period by Myers and Diener (1996). Oswald (1997) found that reported levels of 'satisfaction with life' were only marginally higher than they had been in the mid-1970s. In some countries, including Britain they were actually lower. Meanwhile, in the United States, 'rates of depression have been doubling every decade, suicide is the third most common cause of death among young adults in North America, [and] 15% of Americans have had a clinical anxiety disorder' (Wright, 1995: 53).

What the Green critique adds to this litany of discontent is the notion that, in addition to being both psychologically and sociologically flawed, consumerism is also ecologically unsustainable. Consumption is ultimately the driver for all our energy and material use. Today's levels of material and energy use already threaten potential scarcities in the future; they also place unacceptable burdens on the global atmosphere, on the purity of water supplies, on the integrity and productivity of agricultural topsoils, and on biodiversity. Extending western consumption patterns to the developing world would place even higher burdens on ecological resources. Technological optimists assume that cleaner technologies and more efficient use of resources will provide the basis for the transition to a globally sustainable society. But the impled levels of reduction in material and energy throughput are by no means trivial. Friends of the Earth, for example, argue that the required reductions in the consumption of resources by industrialised nations are in the order of 80–90 per cent (McLaren *et al.*, 1997).

Ultimately, of course, the burden of failure to achieve these reductions will fall on future generations, who may find not only that their own consumption possibilities are severely curtailed by today's obsessive consumption, but also that the stability and integrity of the global commons on which they depend for survival are compromised in potentially catastrophic ways. Thus, the Green critique of consumption assumes an ethical dimension. As Jacobs (1997: 50) points out:

> environmental evidence of excessive resource use then becomes a moral injunction to individuals to consume less. Radical Greens stress each person's responsibility for their own contribution to globally unsustainable consumption.

But the Green critique is not confined to moral imperatives. Some clear attempts have been made to provide alternative models of development, within which quality of life can be characterised in ways which are not bound by conventional economic measures of consumption. After all, as Durning (1992) reminds us, the philosophical and psychological basis of the conventional view – although deeply imbedded in modern institutions – is relatively recent and relatively narrow. A similar argument was made by Fromm (1976), who suggested that modern economic theory finds its philosophical basis in radical hedonism. Though practised through history, particularly by the richest proportion of the population, hedonism was never until recently 'the theory of well-being as expressed by the great Masters of Living'.

Fromm (1976: 4) points to an essential distinction, present in the writings of all those concerned with human well-being, between '(desires) which are only subjectively felt and whose satisfaction leads to momentary pleasure' and objectively valid needs which are 'rooted in human nature and whose realisation is conducive to human growth'. The idea that needs-satisfaction can form the basis of an alternative theory of human well-being has been pursued in more recent ecological writings. Max Neef (1991) has constructed a needs-based theory of development, within which well-being is related to the satisfaction of nine fundamental human needs: subsistence, protection, affection, understanding, participation, idleness, creation, identity and freedom.[6] The Chilean economist goes on to argue that, while the needs themselves are universal in the human psyche, each culture adopts a different set of *satisfiers* in its attempts to meet these needs. Moreover, he points out, these different 'satisfiers' may be more or less successful in meeting the underlying needs. Some kinds of 'satisfiers' may even violate the underlying needs that they are attempting to meet: as an example, Max Neef, cites the arms race as a violator of the need for protection.

Perhaps the most interesting question raised by this framework concerns the relationship between economic goods – between consumption activities (in the conventional model) – and needs-satisfaction. It is fairly clear that this relationship is highly complex, and often nonlinear. More consumption does not always mean more needs-satisfaction. In fact, if the social critiques of consumption are to be believed, it is clear that some at least of the spectrum of economic consumption fails to achieve any needs-satisfaction at all, and may even be violating certain needs (Jackson and Marks, 1999). One commentator has used the metaphor of the 'fat man' to illustrate the point: in a society in which food is scarce and most people are undernourished, the consumption of almost any kind of foodstuff is likely to fulfill the need for subsistence and increase the level of welfare; in a society in which food is more plentiful, the need for subsistence is better met by foodstuffs which have higher nutritional value, rather than by junk-foods; but in a society in which obesity and heart disease are endemic, an increase in the consumption of almost any kind of foodstuff is likely further to violate physical health (protection) in the population and reduce overall levels of well-being (Levett, 2001).

Thus, we are left with the uncomfortable possibility that a certain proportion of economic consumption – with its attendant material throughput and environmental damage – may not be contributing positively to quality of life at all. This situation is clearly more fundamentally dysfunctional than is implied by Tony Blair's opening remarks in the Government's Sustainable Development strategy; nor is it amenable to accounting revisions of the kind envisaged by indices such as the ISEW. Correcting the value of consumption activities for negative environmental or resource effects is not sufficient to inform us what is happening to our quality of life. Some of our consumption not only destroys the environment, it also fails to satisfy our needs and may compromise our quality of life irrespective of its external environmental or social costs.

Of course, it is clear that not all economic consumption has this character. Even with the Max Neef framework, there are certain kinds of needs (subsistence, protection, for example) which demand the provision of certain material goods (foodstuffs, housing, for example). Managing the environmental impacts associated with the satisfaction of these 'material needs' remains a significant challenge. At the same time, it is clear that many of the other needs (affection, participation, understanding, idleness, identity, for example) are 'non-material' in the sense that satisfaction of these needs implies no minimum level of material throughput.[7] Clearly, different cultural 'satisfiers' will have different material implications. For example, attempts in Western culture to satisfy these non-material needs increasingly involve material consumption (Jackson and Marks, 1999). However, if the arguments cited earlier in this paper are to be believed, these attempts to satisfy non-material needs through material consumption are both flawed psychologically and damaging in ecological terms.

Two conclusions follow from this – one of them is stark, the other more hopeful. The stark conclusion is that modern society may be seriously adrift in its pursuit of human well-being. Current consumption patterns not only threaten environmental sustainability, they also fail even to satisfy our needs consistently. The hopeful conclusion rests in the scope for improvement which this perspective offers. Environmental imperatives – the demand to reduce the material impact of human activities – are often portrayed and often perceived as constraining human welfare and threatening our quality of life. In contrast, the radical Green critique suggests that existing patterns of consumption already threaten our quality of life, not just because of their impact on the environment, but also because of their failure to satisfy our needs. Reducing the material profligacy of our lives, according to this view, will help the environment. It also offers the possibility of improving the quality of our lives. 'Revisioning the way we satisfy our non-material needs is not the bitter pill of eco-fascism', argue Jackson and Marks (1999: 439), 'it is the most obvious avenue for renewing human development'.

Concluding remarks

Inevitably, what I have called here the radical Green critique of consumption is not universally accepted even by those with broadly environmentalist leanings. Jacobs (1997: 50) rejects the notion that increased personal consumption fails to improve individual well-being, argues that Greens confuse different notions of quality of life, and accuses environmentalists of political naïvety:

> The radical Green position is barely an argument at all in the sense of a claim addressed to society's collective agency. It is a personal appeal, addressed to individuals and their own sense of individual well-being. And as such it is in a strong sense misdirected.

Jacobs' point is not that Greens are wrong to stress the importance of quality of life. In fact he remains anxious to stress the distinction between quality of life and consumption. He admits that an over-emphasis on private consumption has haunted modern society in recent years and that this 'may actually lead to a decline in well-being rather than to its increase'. The crucial point, in his eyes, is that the Green agenda is essentially a social agenda, environmental costs are social costs, environmental goods are social goods and social goods make a vital contribution to the well-being of society as a whole. But Jacobs view is that we must redefine quality of life in social terms and then appeal to people's interests in the well-being of society before we can hope to persuade them to reduce consumption.

In a curious sense, this position is even more idealistic than the radical Green critique. Jacobs apparent faith in the theory of consumer value leads him to reject the notion that material consumption may be flawed at the individual level. Accordingly we cannot hope to persuade people to forego personal consumption opportunities by pointing to the improvements in personal well-being that may flow from this. Rather we must appeal 'beyond the interests of individuals in their own personal well-being to their interest in the well-being of society as a whole' (Jacobs, 1997: 59). The principal task for environmentalists, Jacobs concludes, is to formulate a concept of quality of life which involves individuals 'as members of society and not simply as autonomous individuals', and to articulate this concept in political terms.

This rejection of the psycho-social critique of consumption implies substantially greater demands on us to reconceive ourselves as social actors than are implied by the Green critique. And, ironically, it implies even greater assumptions about the flexibility of the political system to accomodate this reconception. Today, as Jacobs (1997: 49) himself points out, raising the level of personal consumer spending 'is generally regarded as the key political objective for any government: the principal measure of its success and – according to the conventional wisdom of psephological prediction – the best indicator of its likely vote'. Christie and Warpole (2001: 63) make a similar point:

> Affluent societies breed individualism. Societies such as the UK and the USA have given priority to private affluence and have weakened the financial and human basis of the public realm.

The demand for increased investment in social goods, with the concomitant requirement for higher levels of taxation, is regarded as political anathema in modern democratic market economies.

Thus, the task of articulating a concept of quality of life acceptable in Jacobs' terms is not in itself enough. We must also be able to persuade people to see themselves more clearly as citizens, as social actors, rather than as consumers. Ironically, this task may be significantly harder if we are not allowed to use the argument that material consumption is not necessarily going to make us feel better off–or to be more precise, that foregoing material consumption may lead to a higher personal quality of life. And yet this is precisely the message from critics of consumer culture. The evidence is supportive. The pedigree is strong. For reasons well-known for several millennia, well-being does not consist in the accumulation of material possessions. 'The attitude inherent in consumerism is that of swallowing the whole world', writes Fromm (1976: 27); 'The consumer is the eternal suckling crying out for the bottle.' Philosophers from before the time of the Bible have asked: what profit is there in swallowing the whole world, if you lose your soul?

In reality, the distinction between social quality of life and personal quality of life may not be as clear-cut as Jacobs is suggesting. According to the needs-based model, some of our personal needs have a clear social dimension. Indeed, Max Neef suggests that all of our fundamental needs operate in an interactive or social capacity. Participation, identity, affection all imply social interaction as much as individual striving; even subsistence and protection have interactive dimensions. Nor is this social dimension simply a question of regarding society as a functional source of personal gratification. Our personal identities are intimately linked to our social identities. Our concepts of the good life are not solely driven by concepts of personal gain, but include notions of moral goodness, equity, integrity and justice. And yet these 'moral needs' (Maslow, 1954) are as much personal needs as are our needs for subsistence or freedom. In one sense – a sense admittedly obscured by the individualism of free market economics – appealing to personal well-being is the same thing as appealing to social well-being. This is not to suggest of course that the challenge of appealing to this particular sense of the personal should be underestimated either. Amongst the problems to be faced is the structural role that consumption plays in the neo-classical economy. Irrespective of whether consumption growth improves the quality of our lives, it guarantees the continuity of production, ensures high employment, and keeps the economy from the precipice of recession. 'The alternative to expansion,' as former Tory Prime Minister Edward Heath once remarked, 'is not an England of quiet market towns linked only by trains puffing slowly and peacefully through green meadows. The alternative is slums, dangerous roads, old factories, cramped schools, and stunted lives.'[8]

Of course, a casual observer of British life in the decades since Ted Heath made that remark might have noticed slums aplenty, quite a few dangerous roads, numerous old factories, lots of cramped schools and arguably even some stunted lives. And this would have been in spite of the more or less continued growth that has characterised the economy during those years. Even in their own terms, economies predicated on endless consumption growth are failing to deliver the goods. But this is not quite the point. Whatever the social and psychological dangers of increased material consumption, recession is an even more unpalatable prospect. There is plenty of evidence to show that happiness falls dramatically for those who find themselves unemployed (Oswald, 1997). Economic collapse could presage a humanitarian nightmare.

Here then is the greatest challenge of all: somehow to turn society away from environmental and social dystopia without capsizing the vessel that keeps us from the shark-infested seas. Most realistic suggestions for achieving this task rely heavily on re-establishing the importance of the public realm in government policy: improving public services, defending public goods, protecting community, addressing social exclusion, better regulating the public sector, removing perverse incentives for anti-social behaviour, and articulating new and accessible visions of the common good. In this sense at least, Jacobs is right: Green policy is essentially social policy.

7
The Sustainable Use of Resources

Meg Huby

Introduction

A key way in which Green thinking can inform the development of social policy is by helping to identify situations where meeting the needs of vulnerable people in the short term comes into conflict with the objective of improving longer term social welfare for everyone. Nowhere is this more evident than in the case of ensuring adequate access to domestic energy and water.

This chapter begins by looking at the concerns of social policy, both in relation to the immediate needs of vulnerable groups and to the repercussions which environmental deterioration is likely to have on wider aspects of social welfare and quality of life. It then goes on to examine the notion of sustainable development and what this means for policies to protect social and environmental well-being. We see how dealing with social and environmental problems separately leads to conflicts of interest between short-term and long-term social goals.

Using ideas from across the spectrum of Green thinking, we turn our attention to the dilemma of how to improve access to basic utilities for the poorest sectors of society while reducing energy and water use across the population as a whole. What becomes clear is that, in order to deal with the potentially conflicting objectives of improving both living conditions and environmental quality, integrated policy packages are essential. Single issue policies are simply not enough.

Concerns of social policy

Few people would deny that basic access to light, heat and clean water is a crucial component of well-being. Warmth and clean drinking water are essential for the maintenance of life itself and, at least in western industrialised countries, access to utility services also plays a role in meeting needs in a relative sense. Bradshaw and Hutton (1983: 250) argue that fuel for cooking, light and heating water is required to meet basic needs for everyone. In addition, space heating is a basic need of the very old or sick and the very young. Adequate warmth in the home for the rest of us, if not a basic need, is at least recognised as contributing to normal living standards. Similarly important is access to clean water. Domestic uses of water include not only drinking and cooking but also washing, bathing and the disposal of domestic wastes. Water services in the UK include both supplies of potable water and the removal and safe disposal of sewage, reducing risks from water-borne diseases and benefiting public health as well as individual well-being.

In spite of their obvious social benefits, however, current levels of access to energy and water vary as a result of social and economic inequalities. This is true even within the richer western countries which form the focus of this book. The disparities between these countries and the developing world are far greater and reflect inequities that are similar in nature if not in scale (see Huby, 2001). The problems associated with access to utilities have both social and environmental dimensions.

The social dimension

The need to pay for basic services providing energy and water can reduce the welfare of people living on low incomes, adding to financial hardship by placing additional pressure on already stretched household budgets. Responses to such pressure may include taking on debts, going without other necessities and reducing the use of energy or water to levels which threaten health and hygiene. In extreme cases, failure to pay utility bills has led to service disconnection. Even where services are maintained and debts avoided, considerable distress and anxiety may be caused.

Fuel poverty

In the UK, households needing to spend more than 10 per cent of their income on fuel to maintain a satisfactory heating regime are defined as living in fuel poverty. People aged 60 or more account for about half of all fuel poor households and households with young children about 17 per cent (see Table 7.1).

Even though poorer households spend proportionately more of their income on fuel, many of them still have difficulties in obtaining

Table 7.1   Fuel poverty estimates by household type – all tenures, 1996

Household type	Total households (thousands)	% in fuel poverty
One person aged 60 years or more	3,044	36.3
Couple aged 60 years or more	3,016	13.6
Lone parent with children	1,286	10.5
Couple with children	5,007	6.1
One person under 60 years	2,279	15.9
Couple under 60 without children	3,674	5.5
Other adult households	1,337	12.1
Total	19,643	100.0

Source: Adapted from DETR (1999a), figure E.

sufficient fuel to meet their needs. The problem is particularly marked when age or ill-health confers vulnerability, as shown by a study of people living on low incomes in Britain:

> In winter, homes were bitterly cold. Individual rooms were heated as needed – normally during the coldest part of the late evening. Central heating, even where it was available, was not used. For people who were at home all day, the costs of heating their homes was of particular concern – especially if they needed to stay warm either for health reasons or because they had young children. (Kempson, 1996: 34)

Condensation and damp cause problems in cold houses, encouraging the growth of mould and mites, exacerbating conditions such as asthma and other respiratory complaints. The UK has around 30,000 excess winter deaths each year, many likely to be related to cold living conditions (DETR, 1999a).

Water poverty

Domestic water charges in the England and Wales can represent up to 10 per cent of income for some poorer households, compared to a national average of only one per cent (Office of Water Services, OFWAT, 1996). The consequences of difficulties in paying for water services are rather different from those associated with gas and electricity payments. Charges for domestic energy supplies have a volumetric component which allows customers to reduce bills by cutting down on the amount of energy used. This is also true for the

18 per cent of households with metered water supplies. Metered households can reduce bills by reducing water use but there is concern that health and hygiene may be put at risk if reductions reduce water use below essential levels. Most households still pay for water services on a non-measured basis so, more often, the problems caused by high water service charges are linked to debt and disconnection, significantly more common in low income households (Herbert and Kempson, 1995). In 1989–90, 8426 households in England and Wales had their water supplies disconnected. The figure rose to 18,636 in 1992–93 but subsequently fell, reaching 1907 in 1997–98. (OFWAT, 1999a). Since the Water Industry Act of 1999 disconnection for non-payment of charges and the limiting of supplies to enforce payment has been prohibited, but debt can still be a problem.

People at most risk of getting into debt for water services are among the most vulnerable in society and include those who have a long-standing illness or disability or who have children under five years old. But the extent to which high charges lead to debt does not depend only on the resources available to pay bills. Different water companies have different policies and payment options which help customers to pay their bills more or less easily and these variations often reinforce the inequalities produced by income and family circumstances (Huby and Anthony, 1997).

The environmental dimension

Inadequate access to domestic energy and water causes social hardship but, at the same time, access in itself causes environmental deterioration. Natural resources are consumed in the form of fossil fuels and spatially limited supplies of fresh water while the environment is also used as a sink for emissions of Greenhouse gases and other atmospheric pollutants and for the disposal of waste water.

Greenhouse gas emissions

Human use of energy has grown enormously, based overwhelmingly on burning fossil fuels. This is causing a significant change in the composition of the atmosphere which, unless halted, is likely to have very serious consequences. (Royal Commission on Environmental Pollution, RCEP, 2000: 13)

The problem is that anthropogenic increases in atmospheric greenhouse gases enhance the extent of natural warming of the earth's surface and, consequently, its influence on the oceanic and atmospheric systems which determine climate and drive the world's weather patterns. There is mounting evidence that increasing temperatures are already leading to rising sea levels and increases in the frequency and severity of heat-waves, droughts, floods and storms in many areas. The environmental damage caused by these changes is construed by most Green thinkers as constituting a problem in its own right. But, in the longer term, climate change is likely to have devastating social and economic consequences. In addition to its direct physical effects the impact of rapid change on natural ecosystems it will have knock-on effects on food supplies, human health and welfare. Although exact predictions are dogged by uncertainties, most national governments now take the problem seriously.

Over 170 countries are committed (at least on paper) to international agreements to limit or reduce further greenhouse gas emissions. Following the signing of the Kyoto Protocol in 1997, the UK agreed to a 12.5 per cent reduction in emissions. The Labour Party manifesto in 1997, however, was more ambitious, committing the UK government to a 20 per cent reduction from 1990 levels by 2010. But fossil fuel use, a key contributor to carbon emissions, is still increasing. Meeting reduction objectives will require action relating to the energy supply industry, business, transport, agriculture, forestry, land use and the public sector. But the domestic sector accounts for 30 per cent of UK energy consumption, second only to transport, and households consume a quarter of all UK electricity.

Final domestic energy consumption increased by just less than 25 per cent between 1973 and 1998 but average household consumption fell by about a tenth. Space heating accounts for just over 50 per cent of consumption and water heating for about a third. The decline in the average number of people per household over the last 25 years at least partly explains the fall in average consumption. But at the same time the number of households has increased, as has ownership of items such as refrigerators, freezers, washing machines, driers, dishwashers, televisions and computer equipment. The quantity and proportion of household energy consumption used for lighting and appliances has almost doubled since 1973, accounting for a large part of the total increase (RCEP, 2000).

Water supply

The sources of water for public supply in the UK include rivers, underground aquifers and reservoirs. Although water circulates continuously through the hydrological cycle of the earth and its atmosphere, the stock of fresh water available for use at any one time is extremely limited. Both the exploitation of new water sources and the increasing rates of abstraction from existing sources can have significant detrimental effects. Water plays a crucial role in shaping the natural environment, the levels and flows of surface and ground water in particular areas helping to determine patterns of vegetation growth and the diversity of wildlife habitats. When it is abstracted for public supply or other purposes, the quantity of remaining water resources is lowered. Excessive abstraction leads to low river flows, damaging wildlife habitats and adversely affecting water quality and recreational and amenity value. After use, most water is eventually returned to the environment but the time delay between abstraction and return, geographical differences between points of abstraction and points of return, and the condition of water returned as effluent, all alter the quality of water available for future use.

The need to conserve wildlife and habitats provides a strong reason for protecting the aquatic environment. In addition, any decline in the quantity or quality of water in the environment has implications for its further use as a resource for human beings. Yet water abstraction for public supply has risen steadily over the past 25 years. Between 1971 and 1993, domestic consumption rose by 17 per cent (House of Commons Environment Select Committee, 1996: 428). Water delivered to households is projected to increase by about 1.6 per cent between 1999 and 2000 and 2004 and 2005, largely due to the formation of around 0.7 million new households (OFWAT, 1999b).

As with domestic energy, increases in demand are a function of increasing population, rising household numbers and changes in personal lifestyles, and are expected to continue over the next 20 years (Hills *et al.*, 1997). Lavatory flushing alone accounts for about a third of all domestic consumption (OFWAT, 1999c). It has been predicted that, as a result of demographic and lifestyle changes, the use of water for gardens will increase by 2.5 times on 1991 levels by 2021 and the personal use of water for washing and bathing is expected to rise by 10 per cent (Herrington, 1996: 138).

Sustainable development

The term sustainable development has become common political currency in recent years (see Chapter 1). Agreement about the importance of the concept is found across the political spectrum, from radical bioregionalists to neo-liberal free marketeers. But this consensus overlies a wide divergence of views on how the concept should be understood and employed (Blowers and Glasbergen, 1995; Reid, 1995; Connelly and Smith, 1999). Rigorous Green interpretations of sustainable development recognise the inherent value of the natural world and demand that human activities must take account of this in pursuing social well-being. The interdependence of the human and nonhuman worlds is given a priority that challenges current economic and political paradigms. In contrast, the current dominant interpretation, the one adopted by the UK government, follows the World Commission on Environment and Development in emphasizing the mutual reinforcement of economic growth, social development and environmental protection. It defines sustainable development loosely as 'development that meets the needs of the present without compromising the ability of future generations to meet their own needs' (Brundtland Commission, 1987: 8).

Under this interpretation it is acceptable for current patterns of growth, production and consumption to continue as long as they are modified to reduce their worst environmental effects. Indeed it is argued from the ecological modernist position that increased growth can be used to benefit the environment and that environmental improvement can in turn enhance the potential for growth. This focus on reducing the environmental impacts of activities to produce the goods and services demanded by society wholly fails to question the need for those goods and services. Instead, it perpetuates a capitalist system in which continuing economic growth and consumption are seen as synonymous with social development.

This has implications for policies governing the provision of utility services in that 'the way that individuals consume reflects the way that society values consumption' (Redclift, 1996: 138). It is the consumer culture of western industrial society that has led to the over-consumption of energy and water and so to the associated environmental problems. The impacts of fossil fuel and water use are linked. Greenhouse gas emissions are associated with predicted increases in mean global temperatures of $0.1-0.3°C$ per decade. Sea levels could be 12–67cm higher in 50 years time. In the UK, winter storms will be more severe, rainfall patterns will change and domestic, agricultural and environmental demands for water are likely to increase. It is not surprising then to find that the UK government's set of core indicators for sustainable development includes both levels of household water use and emissions of greenhouse gases (Environment Agency, 2000).

The UK government's interpretation of sustainable development reflects ecomodernist values, stressing the need for continued economic growth. But objectives for sustainable development also imply a form of social progress that recognises the needs of everyone and the need to protect the environment effectively. This suggests that we are trying to sustain two different things. First, we are aiming for sustainable improvements in social welfare to ensure fair and adequate access to domestic energy and water. Secondly, we need sustainable ways to protect and enhance environmental quality, both for its own sake and to ensure that the environment can provide the source and sink resources necessary for continued access to domestic energy and water in the long term. The problem is that short term solutions which see energy and water as commodities (and aim to meet social objectives by increasing consumption) militate against longer-term social welfare by threatening the environmental systems on which continued consumption depends.

Developments in single issue policies

Social protection

If the aim were to address social problems alone, policies could be designed simply to improve access to energy and water services by disadvantaged sectors of society. Since affordability is the main issue here, mechanisms are needed to reduce the price of domestic energy and water services, bringing adequate supplies within the reach of the lowest income groups. Privatization of the utility industries has had an impact on prices across the board. Other policies have targeted low income and vulnerable groups in particular, operating through regulation or by subsidies paid to consumers indirectly via the utility companies or directly through the social security benefits system.

Over the last decade, reforms in the UK energy market, have brought substantial reductions in the price of domestic fuel, reinforced by statutory regulation and helped by a fall in global fuel prices. Increasing consumption of energy is associated with higher living standards, gains in health and welfare and rising longevity, all closely allied to notions of social progress. Nevertheless, the fact that many households still cannot afford to heat their homes adequately has kept fuel poverty high on the UK political agenda. In 1999 the DETR (1999c) published a consultation paper, *Fuel Poverty: the New HEES – a Programme for Warmer, Healthier Homes*. Recognition that fuel poverty results from a combination of low incomes and poorly insulated and heated housing provided the rationale for reforming the existing Home Energy Efficiency Scheme (HEES). One component of the new scheme is New HEES Plus, a package to provide grants for home insulation and heating improvements for certain households in receipt of income or disability related benefits. The maximum grant has been increased from £315 to £700 and savings on heating costs have been estimated at up to £600 a year for a three bed-roomed semi-detached house, depending on the previous condition of the property. Grants to provide materials only have been increased from £160 to £250. The highest number of fuel poor households own their own homes and the private rented sector has the highest proportion of households living in fuel poverty. The New HEES is therefore designed to focus on private properties, leaving local authorities and registered social landlords to take responsibility for the energy efficiency of their housing stock.

Householders have so far been largely exempt from energy taxes. They do not pay the Climate Change Levy exacted from industry, and VAT on domestic fuel was reduced from 8 per cent to 5 per cent in 1997 because the social impact of the tax was regarded as unacceptable. Although presented as a 'Green' measure when it was introduced in 1993, its basis of price rather than carbon content meant that it offered no incentive to switch to cleaner energy supplies. Poorer households spend a higher percentage of their income on fuel so the tax had a disproportionate effect on those with the least capacity to improve their energy efficiency and economise on use. Any environmental benefits of the tax were achieved at the cost of increasing hardship and social inequality. The Spring 1998 budget reduced VAT on the installation of energy-saving materials from 17.5 per cent to 5 per cent for work carried out under government funded schemes, although householders doing their own insulation work must still pay the full rate.

The social security system makes Winter Fuel Payments, worth £100 each winter from the year 2000, available to all pensioners. The Social Fund provides Cold Weather Payments, one-off payments to people considered to be most at risk from the cold. Certain pensioners, people with disabilities and families with young children who live on Income Support or income-based Job Seekers Allowance may claim extra help with their heating expenses during closely defined spells of exceptionally cold weather.

Following privatization in 1989, domestic water prices soared. Between 1989-90 and 1994-95, average household water and sewerage bills rose by 67 per cent (National Consumer Council, 1994). Non-payment could lead to disconnection or limitation of supply and customers had no statutory right to choose to move to measured supplies in order to reduce their bills. The Water Industry Act 1999 introduced new structures for setting water charges and provided new entitlements for household customers. The regulatory system was strengthened and the Director General of Water Services now has the power to approve the charge schemes used by the different water companies. Prices have fallen dramatically as a result of the limits set on the companies and the average household bill for water and sewerage dropped to £219 for 2000–01, a decrease of 12.4 per cent in real terms from the previous year (OFWAT, 2000a).

The Secretary of State now issues guidance to the regulator on the need for protection of specific social groups:

> Charge schemes covering both measured and unmeasured household charges should take account of customers' ability to pay, and address the needs of all those on low incomes, for example through specially designed tariffs and payment options, in recognition of the particular problems which such customers face. (DETR, 1999d: Part 2, para. 6a)

A recent consultation paper proposes that special provision should be made for households in receipt of income related benefits and with three or more children, and households where someone suffers from certain medical conditions. For such households the charges should be capped at the average charge paid by the household customers of each water company for water and sewerage services (DETR, 1999d).

There is little doubt that increased access to the utilities by lower income groups is capable of bringing many social benefits. But increasing access by pricing mechanisms alone has its problems. Making the cost of utilities cheaper gives the wrong signals about the value of natural resources, offers no incentives to conserve energy or water, and consequently leads to greater risks of environmental damage.

As well as the impact of increased consumption on climate change, pollution and habitat disruption, it may well be that, in the longer term, the object of improving living standards for the least well-off is defeated. In the case of water, for example, the establishment of new reservoirs, the transportation of water in response to scarcity and the extension of treatment of drinking water to deal with problems caused by deterioration in quality all increase the costs of water services. As well as higher bills, this also leads to rising costs for other goods and services. Rising food prices may result, partly from the increased costs of water used in agricultural production and food processing, and partly because changes in patterns of natural water quality and quantity affect the kinds of food that can be produced in particular areas.

Environmental protection

In contrast to strictly social policies on the utilities, policies designed solely to protect and enhance the environment require increased prices which reflect the environmental as well as the economic costs of service provision. Higher prices can stimulate efficiency improvements and reduce consumption of both domestic energy and water.

One way to give better price signals for domestic energy is to tax fuel use using a system based on the level of greenhouse gas emissions produced. Carbon taxes would encourage shifts to less carbon intensive fuels as well as increasing the attractiveness of energy saving measures. However, the price elasticity of demand for energy is generally fairly low so the level of taxation would have to be high to bring about reductions in emissions. Figures from the Energy Advisory Panel of the Department of Trade and Industry (DTI) suggest that, to meet the government target of a 20 per cent reduction from 1990 levels, the tax needed would increase the price of domestic gas by 72 per cent and electricity by 23 per cent (RCEP, 2000).

Current plans to reduce greenhouse gas emissions in the domestic sector centre largely on the use of fuels with lower carbon intensity and on increasing the efficiency of domestic energy use. Substantial shifts from coal-generated electricity to gas for domestic heating has already had an impact. Electricity generates about two and a half times as much carbon dioxide as gas for each unit of heat produced. Even better for the environment would be a move away from fossil fuels to renewable energy sources and the Utilities Bill 2000 imposes an obligation on suppliers to supply a specified proportion of electricity from renewable sources. Like taxation policies, the move away from relatively cheap but environmentally damaging fuels will have repercussions on prices: 'Clearly hitting higher renewable targets may, at least in the medium term, mean slightly higher costs to consumers and the acceptability of those higher costs must be fully considered' (DTI, 2000: 7).

Damaging greenhouse gas emissions would also be reduced if households used less energy as a result of improved efficiency. The government's Market Transformation Strategy aims to encourage the production and sale of energy efficient appliances. Efficiency labelling is currently mandatory for refrigerators, washing machines, driers, dishwashers and light bulbs. The Energy Efficiency Standards of Performance Scheme (EESOP), run by the electricity suppliers and

Table 7.2   Potential carbon savings from energy efficiency measures

Possible measure	Savings tC/1000 households	Capital cost £/household
Cavity wall insulation	165–250	250–300
Condensing boilers and controls	230–340	150–250
Compact fluorescent light bulbs	20–25	10–40
Loft and tank insulation	140–210	100–230
Electrical appliances	5–15	0–35

Source: Adapted from DETR (1998) para. e190.

supported by the government's Energy Savings Trust (EST), also helps reduce energy use by requiring the electricity companies to support projects to help domestic customers to improve their efficiency.

Together with the HEES and the Capital Receipts Initiative (CRI) which provides local authorities with funding for improving the energy efficiency of their housing stock, these mechanisms have the potential to make substantial savings in emissions of greenhouse gases by the year 2010 (see Table 7.2).

UK policy on protecting the aquatic environment is partly driven by EU Directives but includes programmes of work to meet domestic statutory obligations and non-statutory work driven by current government environmental policy. The new Water Framework Directive, aimed at achieving good status for surface and ground water, has costs estimated at £4–12 billion. This money must be recovered substantially from water users. It will affect household bills once the Directive is implemented from 2005 to 2006 and it is hoped that increased prices will encourage customers to recognise the true value of water, providing incentives to use water responsibly and avoid waste (OFWAT, 2000b).

Most households pay for their water on an unmeasured basis and have no financial incentive to economise. For others, the kind of tariffs used for water charges can play a key role in environmental protection: 'Tariffs can give strong messages to metered customers to use water efficiently and can also reflect the costs of providing additional water' (OFWAT, 1999c: 23). The Secretary of State's draft guidance calls on water companies to develop and implement new tariffs for metered supplies specifically to encourage conservation and efficient use (DETR, 1999c).

Since the introduction of the Water Industry Act 1999 household customers have the opportunity to choose whether they should remain on unmeasured water tariffs or change to measured water charges.

Compulsory metering is now only applied where households use large quantities of water for non-essential purposes such as filling large ponds or swimming pools or watering gardens with sprinklers. In 1999–2000 only about 18 per cent of households were paying for water on a measured basis. The Director General of Water Services suggests that voluntary metering can only achieve water savings of up to about 5 per cent (OFWAT, 1999b). This is because most households move to metered supplies in order to pay less for the same amount of water. They have little incentive to reduce water use.

Conservation of water can be promoted by companies offering leakage detection and repair services to households and by encouraging the installation and use of water efficient domestic appliances. The DETR has produced regulations governing the efficiency and design of newly installed water fittings. These include a maximum flush volume for water closets of 6.0 litres and the use of dual flush systems, a maximum washing machine volume of 27 litres per kilogram load and, for dishwashers, 4.5 litres per place setting. The volumes are still, however, higher than EC eco-label standards and the regulations do not cover existing appliances. Policies designed to protect and enhance the natural environment are essential for improving and enhancing long term sustainable social welfare as well as for the sake of the natural world in its own right. However, they almost invariably lead to increased costs of water and energy services for domestic customers. It can be argued that these reflect the real costs of provision and that the environmental problems we are now facing are a result of artificially low prices in the past. Environmental protection may require some short term sacrifice of comfort, pleasure and convenience for better off households. The immediate social problem is that higher prices may lead to reduced use of utilities among the less well-off to the extent that risks to health and welfare are increased. It is lower income households who are least likely to be able to afford the capital costs of installing energy and water efficient systems and appliances that would enable them to reduce measured bills in the longer term.

The dilemma

The two types of policy, one focusing on social and one on environmental concerns, have different consequences:

1. Under the policies designed for social protection, some of the poorest and most vulnerable households benefit in the short term.

However, such policies often work to the detriment of the environment and longer term costs accrue that are both environmentally and socially detrimental. When it comes to bearing the brunt of environmental degradation, the poor are likely to suffer most (Huby, 1998).

2. On the other hand, when policies are pursued that prioritise the need to protect the environment, the poor suffer in the short term. In the longer term, however, environmental benefits are enjoyed by everyone, including the poor.

Right across the spectrum of Green ideology lies the recognition that sustaining the quality of the environment is inextricably linked with improving welfare and reducing social inequalities. The goal of sustainability implies that we have some kind of moral responsibility to consider the long term social and environmental impacts of policy. This leads to the choice of (2), above, as the preferred type of development. The problem then becomes one of how to attain the long term environmental goals which promise to benefit everyone, without imposing disproportionately high short term costs on the poorest social groups. Is it possible to reach a point where energy and water use begins a gradual sustained decline without producing unacceptable effects on social welfare, equity and cohesion?

Policy options

People are likely to use less water or energy (a) if it costs more to use more, (b) if they are able to obtain the same benefits from lower use and (c) if they recognise that by doing so they can protect the environment. Let us look at each of these in turn.

Increasing unit prices to reflect external costs through taxation and regulation

We have seen how fuel taxes based on carbon intensity and regulation of tariff structures and charges for water could produce environmental benefits by providing incentives to households to economise on water and energy use. But the demand for both energy and water tends to be price inelastic at basic levels of use (Ernst, 1994; RCEP, 2000) so that if low income households are to meet their basic utility needs they will inevitably face higher bills.

The RCEP (2000: 119) comes down firmly in favour of carbon taxation:

The government is mistaken in keeping domestic fuel cheap for all households in order to help a minority of households who suffer from fuel poverty, when there are growing concerns about the environmental damage caused by indiscriminate, inefficient consumption of fossil fuels.

The House of Commons Environmental Audit Committee (1999) also expressed their belief that the domestic sector should not be permanently exempt from the environmental consequences of its energy consumption but they emphasised the need to address the problem of fuel poverty for both social and environmental reasons. Similarly, the Secretary of State for the Environment acknowledges that the environmental improvements made possible through increased water charges pose problems for low income customers:

> We are concerned to place the supply and regulation of water and sewerage services within the context of our approach to social issues, and social exclusion in particular, and to ensure that sustainable development is taken into account. (DETR, 1999b)

Finding the means to assist vulnerable households raises the question of where responsibilities lie. Should financial subsidies be provided by the social security benefits system, by other utility customers or by the privatized companies?

One suggestion is that, with the introduction of new carbon taxes, the alleviation of fuel poverty could be achieved through the benefits system, funded by ring-fencing a substantial proportion of carbon tax revenue. Levels of mainstream benefits, Cold Weather Payments and Winter Fuel Payments could be increased; outreach systems improved to increase take-up; eligibility widened; and social housing stock improved to increase energy efficiency (RCEP, 2000).

In the electricity industry, the EST oversees the EESOP schemes, which are expected to save 2 million tonnes of carbon emissions a year by 2010 and are funded by a levy on domestic energy. Households all pay the same amount, £1 a year, and 60 per cent of the scheme's expenditure is aimed at low income groups so that the scheme represents a form of cross subsidy: 'The EESOP scheme has been, in effect, a small hypothe-cated and broadly redistributive tax which very few consumers are aware of; it is not mentioned in their bills' (RCEP, 2000: 100).

Cross subsidies to protect vulnerable customers are used by the water companies by means of social tariffs. The water regulator, however, is keen that assistance should be closely targeted and is aware of the impacts of the system for other customers. The Secretary of State's guidance shares this concern:

It is important to recognise that any measure which results in lower bills for some customers will have to be funded by higher bills for others, some of whom will have a relatively low income and may be just above the threshold for assistance. (DETR, 1999c: para. 2.7)

The costs of social tariffs for metered water customers were expected to be less than 50p per household for 2000–01 but will increase as more households take advantage of the scheme and the number of metered households grows (OFWAT, 2000c). It is estimated that up to 55 per cent of pensioners and 47 per cent of households on income related benefits could benefit from changing to metered supplies (DETR, 1999c). The OFWAT National Customer Council (NCC) is critical of direct cross subsidies to other customers:

Our view has always been that Government has the primary responsibility through taxation and the social security system to provide financial support to people on low incomes. This includes help to pay their bills for water and other services. We do not think it is right for customers in general to pay higher bills to subsidise those on low incomes or with special needs as this will mean the movement in bills will be less predictable. The water charging system should not be used to achieve the government's social policy objectives. (OFWAT, 2000b: 14)

Perhaps a more valid criticism of any scheme to give financial help, whether through welfare benefit packages or other forms of subsidy, to help low income households with their utility bills is that it simply allows them to spend more money on what are often extremely inefficient ways of using energy and water. On their own such schemes neither offer incentives to save nor raise awareness of the need for environmental protection.

Promoting more efficient use of energy and water

'Decent, energy efficient homes contribute to social cohesion, improved health and better use of fossil fuels and other resources' (DETR, 1999d). Current attempts are being made to improve the efficiency of new and refurbished housing through the Energy Efficiency Best Practice Programme, offering advice to local authorities and housing associations, the Government Standard Assessment Procedure for home energy rating, and the use of higher standards for building regulations increasing the minimum energy performance of new houses and extensions or conversions. But initiatives aimed at new buildings tend to benefit better off buyers while households experiencing fuel poverty tend to live in the oldest, least efficient housing.

The Home Energy Efficiency Act 1995 does require all local authorities to implement measures to improve the energy efficiency of *all* residential accommodation in their areas but a report to Parliament by the Secretary of State in April 1999 revealed that housing in one in three English council areas showed little or no improvement (Environmental Data Services, ENDS, 1999a). One of the problems is that more than 80 per cent of housing is outside of the direct control of the authorities so they have little scope to set priorities. Increased funding is needed to enable local authorities to repair and improve their existing stock.

The EESOP scheme has led to some efficiency savings but not all of these benefit the environment. The EST estimates that low income households currently take half of their theoretical savings in increased comfort while the figure for better off households is 20 per cent (ENDS, 1999b). Nevertheless the EST has called for an extension of the scheme to include gas and to increase the levy to at least £2. Around 80 per cent of households are now heated by gas and the EST estimates that an extended scheme could treble the savings in carbon emissions. Reform of the HEES has already been announced but it remains to be seen whether the new scheme will address the criticism that it does not target the worst properties. Current spending averages £200 per household and the EST suggests that a properly effective scheme would cost £600 to £800.

At the level of energy supply, combined heat and power systems (CHP) can improve efficiency by making use of the heat produced as a by-product of electricity generation. Even better is energy generated from renewable sources. The EST is to develop a system to promote the marketing of 'Green electricity' tariffs. The percentage components of renewable energy in supplies will be identified with the aim of enabling customers to choose, and to allow environmental groups to promote the soundest options.

Regulations already require that certain domestic appliances, such as washing machines, are labelled to enable people to identify the most water and energy efficient models. Nevertheless, initial purchase costs still put them out of the reach of many low income households. Even for the better off, initial price, appearance and performance may be seen as more important than the longer term financial savings and environmental benefits offered by more efficient products. One suggestion of the RCEP (2000) is that government departments, local authorities, the National Health Service and government agencies should purchase more efficient products in bulk, thus expanding the market and bringing down prices for domestic households. Efficiency devices potentially have an important role to play in conserving water. Trials indicate that simply using 'hippo' bags to displace water in lavatory cisterns and checking for leaks and dripping taps could result in savings of 11 to 15 per cent in the average household. And the costs to water companies of efficiency promotions are considerably less than the costs of developing new water resources (OFWAT, 1999c).

Since 1996 water companies have had a duty to promote the efficient use of water but different companies have different schemes. Some are carrying out trials on water efficient housing, grey water recycling systems and gardening promotions. Some issue free water saving devices and a few are beginning to develop tariffs to encourage water economy. Little is being done so far, however, to share results with other companies to promote best practice (OFWAT, 1999c).

Information and awareness

The two options above are open to the criticism that increased prices may do little to affect the levels of water and energy use in better off households. Indeed, increased efficiency may act to offset the effects of higher prices, thus reducing the number of households for whom price becomes a factor in determining levels of use. It is clear that more is needed to effect changes in behaviour that will protect environmental and social interests for the future. Green ideas challenge the notion that economic self-interest is the only force motivating behaviour and suggest that change can also be promoted by appealing to collective interest. Long term educative programmes are needed to raise and sustain an awareness of the need for sensible use of energy and water to protect the environment and ensure continued supplies. Sustainable development can only be achieved through a major shift in public attitudes and behaviour. Price signals and efficiency policies need to be used as part of an integrated package which includes subsidies incorporating efficiency incentives and credible, easily understood advice and information about the need for change. The EST runs about 50 Energy Advice Centres and the government is promoting an 'Are You Doing Your Bit?' scheme through the media. Yet still the RCEP (2000) notes little public awareness or acceptance of the measures needed to accomplish and sustain reductions in greenhouse gas emissions.

Policies to target young people are essential and could be implemented through schools. The Centre for Sustainable Energy has designed a programme involving young people in the study of energy, both to raise awareness in future energy consumers and to encourage current efficiency improvements. It links the aims of the Home Energy Conservation Act with the requirements of the National Curriculum in a wide range of subject areas. Implemented through local authority education departments the programme provides support and guidance to authorities and teachers to help them market, promote and utilise the teaching materials. Some water companies include work with schools among a range of diverse approaches to encourage water efficiency. Some run media campaigns, visitor centres and promotions at garden centres and agricultural shows. All companies produce advice and information on water efficiency but the regulator has noted some scope for improvement in content and in the way information is distributed (OFWAT, 1999c).

More needs to be done to deliver the message. Sustainability objectives would be better promoted if all information provided on prices, efficiency and energy or water poverty included some explanation of environmental concerns. The HEES documentation could include reference to the environmental as well as the social benefits it offers. And OFWAT could do more to promote the sustainable goals of the Environment Agency, rather than presenting them simply as drivers of price increases. The media have a crucial role to play. Coverage of the announcement of the Utilities Bill in January 2000 clearly militated against environmental understanding by emphasising the aim of tackling fuel poverty and prioritising customer interests by cutting fuel bills.

Political conditions for sustainable development

For behavioural change to take place, people need to feel that their actions make a difference. They need to have a sense of ownership regarding the society and environment in which they live so that they can take responsibility for their own choices about how they use domestic energy and water. This can only happen in conditions where information is readily available and where relationships between citizens and experts are open and informative. Policy-makers need to work with citizens, businesses and non-governmental organisations, encouraging participation in decision-making and ensuring that the policy process is a transparent one. In particular, the problems posed by scientific uncertainty about the risks of environmental degradation need to be treated with more candour. On the positive side, both the

Water Industry Act 1999 and the Utilities Bill 2000 were preceded by extensive consultation processes, with documentation freely available on departmental websites.

Privatization of the utilities has done little to help the cause of sustainable development. The 1989 privatization of water was, at least in part, presented as a 'Green' strategy. The government claimed that it would not only benefit domestic customers but would establish a clearer framework for environmental protection (Saunders and Harris, 1994). It is, however, questionable whether private water companies, geared to maximizing profits and share dividends, are capable of delivering a national plan for the protection and management of water resources to benefit both the environment and current and future generations (Middleton and Saunders, 1997). Market liberalisation and falling prices certainly seem to be at odds with efforts to promote water and energy conservation.

The opening up of the energy markets in 1999 gave households the opportunity to choose their suppliers. But competition on the basis of unit price has led to lower prices and free offers to attract customers. Together with the objectives of industry regulators to reduce costs to households, it leaves little incentive for efficient energy use. Consumers are likely to change suppliers to cut bills rather than improve their energy efficiency. Four companies even offer reduced rate tariffs for consumption above a certain level. Not only does this encourage profligate use but the higher unit rates at low consumption levels discriminate against low consumption households (ENDS, 1999c).

Ernst (1994) argues that utility privatization has been built around an organising principle of consumerism that is inconsistent with the provision of essential services like domestic energy and water. Water and energy have a unique status in that up to a certain level of consumption they are essential and non-substitutable. Their contribution to public health and social well-being characterises them as merit goods. Taken together, these qualities confer the status of social goods and give the government a proper interest in the way that energy and water services are provided to meet social and environmental objectives (Hills *et al.*, 1997). It is only beyond the level of essential consumption that they can be treated as commodities, subject to market forces. Many of the problems besetting the drive for efficiency and social and environmental protection can be related to the way that market values have been attached to energy and water at all consumption levels in a society that currently sees high consumption as an indicator of high living standards. Indeed, the capacity to waste energy and water may be seen by some as a sign of affluence and status (Bhatti, 1996).

This context presents difficulties for the utility industries and government in balancing pressures to reduce bills and pressures to deliver environmental improvements. Markou and Waddams Price (1997) argue the need for a framework within which the privatized industries can work effectively but which clarifies where responsibilities should lie for the social and environmental impacts of utility provision. Such a framework must allow for the balancing of interests between customer bills and service levels, environmental improvements, returns to the utility companies and levels of investment needed to maintain the industries' assets.

One way forward might be to harmonize the, often conflicting, objectives of the industries, government departments and their agencies under the umbrella of sustainable development. The unique status of energy and water, as social goods as well as commodities, means that decisions about their provision cannot be left solely to market forces. Government has a very proper interest in the matter. Through the industry regulators the government already takes an interest in the level of charges to households. But as a steward of the environment it has an interest in ensuring that there exist incentives for the sensible use of energy and water, by both the industries and domestic consumers. For reasons of social justice it also has an interest in the way that the costs of provision are distributed between households.

Although nationalisation of the energy and water industries is unlikely to be re-introduced in the near future, there is a crucial role for government, acting through stronger regulation, to lead the industries towards strategies that command political and public support. But public confidence in the regulatory system is only likely to be gained if the regulatory authorities involve in debate and decision-making all of the stakeholders they represent. The privatisation programme has illustrated that, rather than leading to the withering away of the state, the mechanisms of public policy need to remain centrally engaged in public utility provision if vulnerable consumers are to be protected. And, as we have seen, social protection in the long term cannot be divorced from the protection and enhancement of the natural environment.

8
With No Strings Attached? Basic Income and the Greening of Security

Tony Fitzpatrick

Introduction

The profile of the reform proposal known as Basic Income (BI) has risen considerably in recent years.[1] Although the rationale behind the idea has existed for several centuries (see below) only in the 1970s and 1980s did it begin to be considered seriously by academics, pressure groups, political parties, trade unions and governments. Indeed, the literature dealing with BI is now so vast and, often, so sophisticated that several books would be required for an adequate summary. By the late 1990s BI was being seriously considered by the Irish government and several countries, e.g. Netherlands and Brazil, although ostensibly pointing in another direction, seemed to be backing towards a BI system. In fact, a few voices within the New Labour circle were suggesting that the UK was also doing so and that as tax credits become more and more important within the benefit system the UK will be left with a BI scheme in all but name (Jordan *et al.*, 2000).

What follows is, first, a summary of the BI proposal and, second, an attempt to highlight the pros and cons of the proposal for ecologists. A much more extensive and detailed treatment of the subject can be found in Fitzpatrick (1999a).

A summary of Basic Income

Although BI is, first and foremost, a proposal to reform the tax and benefit systems it has wide implications for economic, employment and social policies in general. A BI would be received by every man, women and child periodically (whether weekly, monthly or annually) as an unconditional right of citizenship, i.e. without reference to marital or employment status, employment history or intention to seek employment. It would replace most of the benefits, tax reliefs and tax allowances that currently exist, and could be age-related, e.g. with a higher BI for elderly people. BI therefore potentially represents an alternative both to means-testing and to the social insurance principle, although, as we shall see, some insist that it overlaps considerably with these.

The long history of the proposal (Fitzpatrick, 1999a: 40–4) stretches back to Tom Paine in the 1790s. Paine had already sketched the outline of a welfare state in *The Rights of Man* and, several years later, went on to recommend the creation of a national fund out of which a lump-sum grant would be paid. Such ideas continued to bubble beneath the political surface for several generations, erupting again into life in the 1920s and 1930s largely due to the work of C. H. Douglas whose influence was especially pronounced in Canada. As the classic welfare state was founded in the 1940s some, such as Juliet Rhys-Williams and James Meade, argued that it should either be supplemented or replaced by a BI. Indeed, by the 1960s both the Labour and Conservative parties were giving serious consideration to variants on the BI proposal, although more traditional ideas were eventually to prevail. As already indicated, BI surfaced in a number of different forms in countries such as Belgium, Denmark, Netherlands and Canada in the 1970s and, by the mid-1980s, this level and diversity of interest began to be co-ordinated through the Basic Income European Network. By the late 1990s this title had made itself redundant as the network had long since become truly global in scope.

As such, the question as to how generous a BI would or would not be varies from nation to nation (e.g. Fitzpatrick, 1999a: 38–40). However, all agree on the following terminology. A partial BI would not, by itself, be enough for an individual to live on and would need to be supplemented by income from other sources, wages for instance. It would, though, be largely revenue neutral. That is, it could be afforded by transforming and re-directing expenditure on the existing tax and benefit systems. A full BI would be enough to live on, but would require additional levels of expenditure for it to be introduced. As we shall see below, this distinction between a partial and full BI is a source of both potential weakness and strength.

BI has been defended from all parts of the ideological spectrum, although for different reasons and with different implications for its specific design and implementation. For those on the Right, it is the logical extension of a means-tested system, a way of systematising all means-tested provision so that a universal safety-net is effectively hung beneath the income ladder. Therefore, some argue that BI is little more than a Negative Income Tax where income, or 'negative tax', is paid to those who fall below a stipulated level (Fitzpatrick, 1999a: ch. 5). Those in the political Centre wish to emphasise the continuities between BI and the existing benefit system. Tony Atkinson, for instance, argues that a BI would need to be constructed upon the social insurance principle, plugging the gaps that insurance benefits traditionally left unfilled. So, a Participation Income would not be received unconditionally but only by those who perform some socially useful activity, albeit activity that is not limited to employment (Fitzpatrick, 1999a: ch. 6). For those on the Left, BI only makes sense as a precursor to a social dividend scheme, i.e. one where the unconditional income is funded out of the value generated by collectively-owned resources rather than income tax *per se* (Fitzpatrick, 1999a: ch. 7). In addition, feminists have also supported the idea, as have many who are less easy to locate in terms of the political spectrum. Ackerman and Alstott (1999), for example, defend a one off lump-sum grant of $80,000 that would be paid to every young adult, a libertarian idea intended to appeal across the political spectrum but one that would lead to a stakeholding welfare system. In practice, the difference between such a grant and BI depends upon the frequency with which the latter would be paid out (Ackerman and Alstott, 1999: 210–16).

Despite this political variety the arguments for and against BI centre upon the following points (Fitzpatrick, 1999a: ch. 4). Its supporters claim, first, that a BI would be more effective than existing policies at guaranteeing a minimum income for all. Whereas 4–5 million people currently fall below the means-tested minimum in the UK a BI would establish a floor upon which all could reasonably be expected to stand since, being unconditional and universal, take-up would be close to 100 per cent. Secondly, BI would embody the equal rights of all citizens. Third, it would tackle poverty and unemployment traps for, be being unconditional and therefore constant, it would not be withdrawn as people either move into employment or experience an increase in their earnings. Fourth, it would improve the freedom and security of individuals, particularly important in the flexible and insecure labour markets of today's global economy. Finally, it would be easy to understand (few have a complete grasp of existing benefit and taxation systems) and cheap to administer.

However, BI's critics allege the following. First, that it would undermine an ethic of citizenship by being unconditional and therefore ignoring the necessity of duties and responsibilities. Citizenship is not only about status but also about participation and active membership in one's social way of life. BI, though, reflects a rights-based view of social membership and, by not requiring anything in return, would encourage individuals to take without giving. Second, a full BI would be too expensive, requiring a level of taxation that is not politically feasible. A partial BI would be affordable but, since it would not provide enough to live on, would be less effective at tackling poverty than the present benefit system. (It should be noted that this criticism depends upon a static interpretation of taxes and benefits: although a full BI would require higher taxation, research suggests that it would nevertheless redistribute to those on lower and middle incomes; a partial BI would be easier to supplement than existing benefits due to the high withdrawal rates of the latter.) Finally, BI fails to attract enough political and popular support because although it has many attractions none, by itself, carries enough political weight.

Where do Green ideas fit into the debate? To what extent can a BI be thought of as a Green social policy? Should ecologists be supporting the idea or not? Before addressing those questions (see Fitzpatrick, 1999a: ch. 9) it is worth understanding why ecologists object to the existing model of social security and where Green political parties currently stand *vis-à-vis* BI. Doing so is the aim of the next two sections.

Social security

The Green critique of social security follows on from the general critique of the welfare state that was offered in Chapter 1, i.e. that full-time, full employment is neither sustainable nor desirable. Ecologists recognize that the modern system of taxes and transfers provides people with a relative freedom from the labour market, but they point out that it does so only by tying them into that market at a much more fundamental level. In other words, the decommodification offered by system is premised upon an overarching commodification of social activity and communal identity. Yet, even in terms of this limited objective, benefits (whether assistance or insurance) fail to guarantee a minimum income for all. The Green critique therefore consists of two components: a conventional criticism of the benefit system's failures, focusing upon the considerable holes in the safety-net as well as its propensity to create unemployment and poverty traps; a specific ecological criticism, focusing upon the extent to which the system is organised around the employment ethic and, by extension, around the growth-oriented productivism of industrial capitalism.

At first glance these criticisms may appear contradictory. The benefit system is charged both with making employment more difficult and with being based too heavily upon the norm of waged labour. However, the apparent contradiction clears once we remind ourselves that ecologists are concerned both with social justice, i.e. with improving the welfare of the least well-off, and with environmental sustainability. Their aim is not to abolish paid employment either as a source of income or well-being: indeed, Greens accept that a major cause of poverty is the fact that the poor do not have enough employment because jobs, at least the best-paid jobs, are concentrated upon a fortunate minority. Instead, the aim which Greens recommend concerns sharing out existing employment-levels more fairly: social justice is not so much about increasing employment-levels and, by implication, ecologically insensitive growth, as about freezing GDP growth at its current level and redistributing the jobs that already exist. Thus, in order to share employment more evenly the emphasis must be taken *away* from the employment ethic (Fitzpatrick, 1998). Rather than imagining that there can be full-time, continuous employment for all who want it we should make jobs less central to our lives.

In short, many ecologists focus upon the necessity and the desirability of working-time reductions (see Chapter 10) and a brief calculation will show why. In a society of 100 people where 90 have jobs and 10 are unemployed, should we (a) attempt to create jobs for the unemployed 10, or (b) redistribute those 90 jobs? According to ecologists strategy (a) is no longer viable since it implies 'going for growth' and is an approach that was only partially successful in the period of high industrialism and post-war recovery (1945–70). This leaves (b). Therefore, if those 90 in employment each work 40 hours per week then this means that 3600 hours (90 × 40) are worked in total. But instead of having 90 people work 40 hours a week would it not make more sense to have 100 people work 36 hours per week? This would leave the same employment level of 3600 hours but would redistribute it fairly across the population. In short, jobs can be redistributed by making wage-earning less central to individuals' welfare. It is here that the significance of BI becomes visible as a means of facilitating this change, as a 'second cheque' that will compensate for loss of earnings and as a way of encouraging people to engage in non-employment forms of work.

Like the welfare state in general, then, the social security system is charged with being blind to the environmental consequences of growth, by being based upon the productivist employment ethic, and with failing to guarantee a minimum income for all. The ecological alternative would make it easier and more worthwhile for people to reduce the numbers of hours they work by taking the emphasis away from full-time continuous employment.

Green support for Basic Income

BI is usually treated as the archetypal Green social policy, though quantifiable evidence to this effect is very hard to find. In the summer of 2000 I attempted a worldwide survey of as many Green political parties whose addresses I could find, asking whether or not they supported BI (or a similar alternative) and why. The results were disappointing to say the least. Out of about 50 parties contacted only a few replied with any useful information, although several others signalled their interest in the idea.[2] Broadly speaking, two reasons were given for supporting BI. The first of these relates to Green ideas in particular and the expectation that a BI would assist the informal economy of unpaid labour (volunteering and caring) by making people less dependent upon the wage contract, by making more free time available and so encouraging a greater degree of communal self-reliance. Such has been the long-term position of the British Green party (1995, 1997). Second, respondents mentioned most of the points that have been outlined above, e.g. that a BI would improve individual freedom and security by comparison with existing employment, benefit and tax policies. These are reasons that are not specific to Green ideas, although respondents clearly anticipated that BI would help to shift social activity in an environmentally-benign direction. Whether or not such support would survive the transition to national government is, of course, not something that can be answered with any degree of certainty (the German Greens provide little help in this respect, being one of those parties who have always been lukewarm towards BI).

In short, only an anecdotal assessment of Green party support for BI can be made at this time. Experience suggests that most Green parties support the idea, with varying degrees of analysis, but that an extended examination of BI and ecologism is required. The rest of this chapter provides the outline of such an examination.

Advantages and disadvantages

Ecologists are therefore widely seen as the most vociferous and long-standing supporters of a BI. To see why this is we need to outline the three potential advantages that a BI holds for Green social and welfare reforms. The first of these relates to the potential implications of BI for GDP growth.

According to ecologists, the drive for ever higher levels of economic growth is fuelled by a number of factors: capital's need to accumulate, individual materialism and a productivist interpretation of redistribution (which says that the poor should be helped by increasing the nation's total stock of wealth rather than zero-sum redistribution from rich to poor). Whatever the motivation the common belief is that at a GDP of, say, £800 billion a country such as Britain is twice as well off as it was when GDP was £400 billion. What this economistic definition of welfare assumes, of course, is that the more 'goods' there are then the higher the level of needs and desire fulfillment; but even putting aside the point made long ago by Fred Hirsch (1976), that material goods are 'positional', what this economism ignores is the fact that economic growth also generates 'bads' which neutralize the positive effects of goods. Indeed, Ulrich Beck (1992) insists that a post-industrial society is that which is characterized by its generation of bads. The contradiction which emerges between consumption (good) and the non-consumption of the poorest (bad) leads to numerous social traumas such as crime, hopelessness and addiction, yet both sides of the equation derive from the simplistic link which is made between well-being and the unlimited expansion of material wealth.

According to ecologists, what we need is some way of breaking this link and of slowing the growth process down. A number of proposals have been put forward (see Chapter 6) for changing accounting procedures so that indicators such as GDP are abandoned and ecologically sensitive indicators, which take account of both social and environmental bads, are introduced (Daly and Cobb, 1990; Anderson, 1991; Jackson and Marks, 1994). We would then have a better measuring-rod for assessing the true state of social welfare. But according to some a BI could also help to break this link (Johnson, 1973; Powell, 1989; Daly and Cobb, 1990; Hoogendijk, 1991; Lerner, 1994; Offe, 1996: 209–10). At present, most benefits are distributed to most individuals on the basis of the past, present or potential contributions which they have made, are making or could make to GDP growth. The insurance/assistance model of benefits is based upon this notion of productive contributions and when people defend the conditionality of benefit provision they do so on the basis that 'the world doesn't owe anyone a living' or some such opinion. By being unconditional a BI breaks this link between making a contribution and receiving a benefit and so undermines the rationale of GDP growth. Or, to put it another way, by being paid irrespective of job record and status a BI undercuts the employment ethic and so challenges the productivist assumptions which legitimate that ethic.

Therefore, what other ideologies usually treat as a disadvantage ecologism tends to treat as a virtue (Mellor, 1992: 206). For ecologists, people should be opting out of the labour market: the fewer people that are actually contributing to GDP growth then the more the brakes will be applied to such growth. In fact, we should aim at a full BI as soon as possible in order to provide people with the incentive to abandon wage-earning. In theory, the higher the BI then the slower the rate of GDP growth.

One objection to all of this, of course, is to point out that the full BI envisaged by ecologists is impossible: because as a full BI encourages a mass migration from the labour market then society's ability to fund that BI is correspondingly undermined (Irvine and Ponton, 1988: 73; Kemp and Wall, 1990: 78). However, putting aside the importance which ecologists attach to land and energy taxation as a source of BI funding (see below and Chapter 12), Green BI supporters usually defend an 'optimal BI': one which slows growth down to an environmentally sustainable extent without the economy having to contract. This is illustrated in Figure 8.1.

As the level of BI increases so GDP growth begins to slow (line *X–O*). At point *OP* on the curve BI has reached its optimum level, so that any attempt to slow growth further (line *O–Y*) would only lower the BI level by reducing the tax yield needed to fund it. This lower BI would then encourage people to re-enter the labour market and so create more economic growth. In theory, then, using BI as an anti-growth measure means that GDP growth can only slow to point *O* on the horizontal axis because BI begins to decrease in value beyond point *OP* on the curve.

There are obviously important questions as to what level of income this optimal BI would imply. If it were still a relatively modest level then no great market exodus would occur and GDP growth would not slow down to any great extent after all, i.e. GDP growth at point *O* in Figure 8.1 would not represent a significant change to that at point *X*; thus, a central Green justification of BI would be weakened. Green supporters, however, still insist that a BI is likely to produce the desired

Figure 8.1   BI and GDP growth

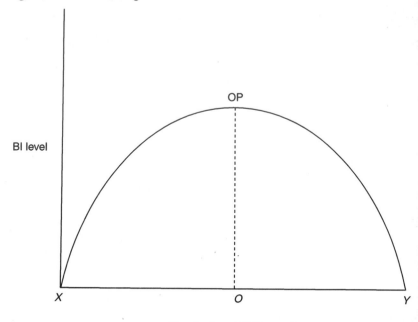

OP

BI level

X   O   Y

The slowing of GDP growth

benefits even if an assessment of its full impact cannot be made until
after it has been introduced.

The second major reason why ecologists support a BI deals not with
its effects on the economy but with its embodiment of an ethic of
common ownership. Tom Paine defended his early form of BI by
arguing that some form of compensation was needed for the way in
which the economic system of private ownership (which Paine
defended) has expropriated that which is, properly speaking, the
common ownership of all. Such compensation would constitute the
basis of the national fund out of which Paine advocated a lump-sum
grant be paid. This argument chimes with the insistence of Greens that
the Earth's resources be thought of as held in common, so that being a
global citizen means being a steward or a trustee whose duty it is to
hand on the Earth to the next generation of common owners. This
does not imply returning the Earth to a 'state of nature' since
resources, especially renewable ones, can be utilised; however, it also
means that a certain proportion of the wealth which ultimately derives
from this common ownership should be shared out unconditionally:

common ownership implies a fundamental *equality* of ownership. Whereas the present transfer system provides most to those who contribute the most to environmentally damaging growth, a BI would express and embody a communal egalitarianism which is not captured by the insurance/assistance model (Achterberg, 1999). Obviously, a BI is not the only means of embodying this communal egalitarianism, but it is one which offers a realistic alternative to the present ways of distributing income by embodying an ethic of common ownership, our common interests and the common good (Jordan, 1992).

The final reason why many Greens support a BI has already been mentioned above (Ekins, 1986; Irvine and Ponton, 1988: 70; British Green Party, 1997; Lambert, 1997: 60–1). Because a BI could reduce or eliminate the poverty and unemployment traps it would make part-time work and, depending upon its level, low-paid work more attractive and so could go some way to achieving the aim set out in the previous section: that of redistributing available jobs by taking the emphasis away from the necessity of working full-time for several decades (Kemp and Wall, 1990: 77). As part-time employment became more financially worthwhile then many people might take up the option of working part-time and it is possible that an expansion in work-sharing activities could result, as well as an increase in people taking sabbaticals from employment. In short, and unlike insurance contributions, a BI: would provide a guaranteed minimum income for all, by repairing the holes in the benefit safety net; would not be a tax on jobs; would improve the range of available alternatives; would permit flexibility without being charter for low-paid casualization.

However, there are also three principal reasons why Greens might not wish to support a BI (cf. Groot and van Iperen, 1999).

First, although a BI could have a role to play in a future ecological society its ability to carry us forward into that society is either limited or non-existent. Mary Mellor (1992: 206–7), for instance, is attracted to the principle of a BI but insists that its introduction is dependent upon a prior change in the ownership of productive property. Looking again at Figure 8.1, the allegation is that the difference between X and 0 is not that great: in other words, the optimum level of a BI would not slow GDP growth down to any great extent. The achievement of a Green society and economy is going to require a massive change in public consciousness and institutional reorganization, whereas a BI might do no more than consolidate existing values, assumptions and habits. Therefore, ecological critics point out that a BI society would be substantially the same as the pre-BI society. According to Andrew

Dobson (2000a), a BI looks like a social-democratic measure grafted unsustainably onto the ailing post-industrial body politic, rather than a radically Green measure in the spirit of solutions to the problems of sustainability raised by the spectre of limits to growth.

Dobson therefore identifies what he takes to be a contradiction between the anti-materialism of Green thought and the fact that BI is dependent upon a high level of material wealth for its financing.

This attempt to disassociate BI with the aim of slowing economic growth down is, indeed, one with which some BI supporters agree. Van Parijs (1992: 26–8; cf. Van Parijs, 1991) argues that far from being anti-growth a BI would be growth friendly due to its positive effects on economic efficiency; if Greens are to support it on economic grounds, therefore, they should do so because a BI might enable growth to be channelled in ecologically friendly directions (cf. Johnson, 1973: 180–9). By contrast, Claus Offe attempts to steer a course between these positions:

> while the right to income as an unconditional citizen right would certainly not by itself alleviate the environmental and ecological risks and dangers of industrial growth and the full employment which is contingent upon such growth, it would probably contribute in indirect ways, for it removes some of the productivist pressures and anxieties and thus paves the political road towards targeted and selective environmental policies, some of which are bound to entail the very termination of certain lines of production and production processes. The Basic Income makes an ecological critique of industrialism politically more affordable. (Offe, 1993: 230)

Or, as the German Green Thomas Schmidt put it: a BI 'seems to achieve very little and at the same time a great deal' (quoted in Hulsberg, 1985: 12).

The second Green objection concerns the unconditional nature of a BI. On one level it is true that a BI would enable people to opt out of the 'employment society', with its ethos of competition and acquisition, and pursue other activities. The trouble is that there is no guarantee that these other activities will be environmentally friendly: perhaps they would merely be parasitic on the very productivism and materialism which Greens strenuously oppose. André Gorz (1992), for example, argues that although a guaranteed minimum income is a necessary condition for social participation it is not a sufficient condition and, until recently, Gorz insisted that the provision of a BI to an individual should be conditional upon the performance of a minimum amount of work during his/her lifetime (see Chapter 10). Alain Lipietz agrees, arguing that an unconditional income scheme is not only unfair on those who finance it, but is morally debilitating for those who cannot find employment and so become dependent upon it:

> a universal allowance of around two thirds of minimum wage would be acceptable only if it meant that those who received it were *prepared* to show their solidarity with society. ... A permanent state of affairs where some people are helped to do nothing is not only provocative to people in work, it is also psychologically painful for those being helped. (Lipietz, 1992: 99)

In response to these kinds of objections, Jan Otto Andersson (1996), has argued that raising the level of an unconditional income in the hope of achieving a full BI is unlikely to encourage those activities and lifestyles which are most conducive to an ecological society. As an alternative, he believes that a small but unconditional BI for all (which can still be supported due to its promotion of personal autonomy) be combined with a Citizen's Wage which, not being a million miles away from a Participation Income, would be provided to those who engage in activity outside the labour market which is judged to be socially useful and environmentally friendly. The objective here is to promote the expansion of the Third Sector, i.e. that which is organized around neither the state nor the market, a goal which Andersson shares with Jeremy Rifkin (1995: 256–67) and now Ulrich Beck (2000) who have also argued in favour of tying the guaranteed income into service in the communal activities of a 'social economy' (see Chapter 9).

Elizabeth Anderson (2000) goes slightly further, arguing that BI's legitimation problem ('why should the undeserving receive a BI'?) could be solved by retaining its unconditionality but tying this into a new system of property ownership. If the resources that we can all be said to own were to be taxed then objectors to an unearned income become less vocal – as is the case in Alaska due to its social dividend scheme (Fitzpatrick, 1999a: ch. 7).

The final objection states that there is a contradiction between the decentralization which Greens desire and the fact that a BI would have to be administered centrally, e.g. to facilitate redistribution from rich areas to poor ones (Frankel, 1987). This objection, though, depends upon a vision of a Green society as being as decentralized as administratively possible, perhaps along the lines suggested by Bookchin (1972) or Sale (1985). However, if we were not to favour this type of anarchistic approach, and if we envisage a continuing role for central government, which seems necessary since a global economy does not cease being environmentally destructive just because a group of people form themselves into a 'bioregion', then the contradiction vanishes: BI can be a form of central administration whose function, in a Green society, could be to facilitate the decentralised self-management of Andersson's Third Sector.

In summary, Greens have various reasons both to support and to oppose a BI. It embodies a social right not to engage in productivist, wage-earning activity (while making it easier to take part-time work) but says little about the obligations which we owe to non-humans and to future generations. It perhaps takes into account the environmental consequences of growth but the effects which its introduction would have in slowing GDP growth down are uncertain. It guarantees a minimum income for all and challenges the employment ethic, but it also seems to depend upon the very ecologically damaging activities to which Greens object. The question, then, is whether a BI could be made to work in such a way that the above objections to it can be discarded. What is the Green 'policy package' within which a BI could be made to work most effectively towards environmental objectives?

Beyond Basic Income?

There are aspects of Green support for BI with which I am uneasy. One option would be to provide a BI for the first two children only and not for any subsequent children. The rationale at work here being to bear down on population growth by providing people with a disincentive against having too many offspring, i.e. to pass the full costs of additional children back to the parent(s) involved. This kind of reform has been advocated by pro-BI ecologists such as Johnson (1973) and Irvine and Ponton (1988: 70).

My own inclination is to reject any such suggestion. The populations of developed countries are relatively stable and there is no evidence that universal benefits make people more promiscuous and therefore no reason to anticipate that a BI would do so. However, what if a BI were to be introduced into a country where population growth *is* a problem? Might it then make sense to employ the policy as a 'social contraceptive'? Again, I believe not. This is not to claim that population growth is not a problem – although ecologism has an unhealthy history of getting into a

'Malthusian panic' over the subject which can shade into racism – but to claim that the merits of birth control are best promoted through educational measures rather than through restricting a benefit to children, many of whom are likely to suffer as a result.

Therefore, we are best thinking of a BI has part of a Green policy package. There are three other policies which sit comfortably within that package: informal economics, working-time reductions and eco-taxation. Each of these are dealt with elsewhere throughout this book (in, respectively, Chapters 9, 10 and 11) and I have little to add to what is said there (see Fitzpatrick, 1999a: 192–201). However, I do want to say something specifically about informal economies as they have been examined by Claus Offe. Like Gorz, Offe arrived at ecological ideas via Marxism and may be defined as a leading theorist of ecosocialism. Also like Gorz, Offe's work is wide-ranging and varied (Pierson, 1991: 40–68) but here we shall only need to concentrate upon the following: the role of welfare services under advanced capitalism, the significance of new social movements and alternative forms of welfare provision.

It has often been pointed out how the neo-Marxist critique of postwar capitalism given by Habermas, Offe *et al.* resembles that of the radical Right. Both ideologies identify a conflict between capitalism's need to maintain the conditions of capital accumulation and the welfare state's tendency to undermine those conditions. But whereas the radical right insisted that the solution was to abolish, privatise, marketise or residualise the welfare state (depending upon who you read), neo-Marxist commentators saw this as an example of pro-capitalists misunderstanding the nature of capitalism: for capitalism requires not only 'accumulation' but also 'legitimation' (cf. J. O'Connor, 1973). Using a 'systems analysis' Habermas (1975) and Offe (1984) identify certain 'crisis tendencies' to which capitalism is said to be prone.

The political–administrative system 'steers' and regulates the economic system and so prevents it from malfunctioning; in return the economic system provides the state with revenue. The state uses this revenue to finance state welfare systems which secure mass loyalty to state capitalism partly by enforcing its imperatives and partly by ameliorating its worst effects. The conflict between capitalist economics and state welfare, therefore, is not due to the misguided actions of postwar politicians but is due to the inherent contradictions of capitalist subsystems. Social policy is therefore an element in the self-regulatory mechanisms of 'organized' capitalism which always threatens to collapse into a series of crises: economic, political and socio-cultural. This analysis of the welfare state is famously summarised by Offe:

> The embarrassing secret of the welfare state is that, while its impact
> upon capitalist accumulation may well become destructive ... its
> abolition would be plainly disruptive. The contradiction is that
> while capitalism cannot coexist with the welfare state, neither can it
> exist without the welfare state. (Offe, 1984: 153)

Subsequently, Offe (1985, 1996: 147–82) insisted that organized
capitalism was giving way to a phase of *disorganised* capitalism where
mass production, state intervention, full employment in the context
of a mixed economy, centralised administration and the post-war
stand-off between capital and labour were giving way to newly
emerging forms of social power and economic organisation. What
this implies is the gradual seeping away of support for state welfare as
the ability of the state to integrate social groups and classes is seen to
decline due to the increasing hegemony of global capital. The
Keynesian welfare state becomes a memory along with the mixed
economy of full employment upon which it depended. This does not
necessarily lead to the disappearance of state welfare, as most on the
radical right have both desired and envisaged, but it does mean that
welfare systems become more authoritarian and residualist as citizen-
consumers are expected to take greater responsibility for their own
well-being. Jessop (1994, 2002) has referred to this as the workfare
state.

For anyone worried about these trends the obvious question to ask
is: where do we go from here? Like many on the Left Offe has
speculated about the possibility of a 'coalition' between the Left's
traditional constituency and the new social movements, e.g. animal
rights, civil rights, ecological, gay rights, peace and women's
movements. Potentially, such movements are as socially significant as
the bourgeois movements of the eighteenth century and the workers'
movements of the nineteenth and early twentieth centuries (Offe,
1996: 19). The obvious problem is that these movements have not
established, and perhaps cannot establish, a distinctive and commonly
shared political programme. One of Offe's recent objectives, therefore,
is to trace the possible contours of such a programme and so to sketch
the outlines of a welfare system which could succeed both the
Keynesian welfare state of organized capitalism and the workfare state
of disorganized capitalism. It is when this aspect of his work is
considered that Offe can be described as an ecosocialist.

Offe's (1993: 74–7; 1996) support for a BI is somewhat cautious. He is
wary of overstating the potential of a BI to inaugurate a future order of

social justice; instead, he thinks that it might make more sense to stress BI's capacity to defend the concept of social justice while adapting it to our existing social environment. Therefore, Offe's defence of a BI is circumspect and he tends to regard it as having a radical potential only if it is included as part of what he himself calls a policy package, the aim of which should be to encourage the growth of informal economies and 'co-operation circles':

> our model of a 'co-operation circle' proposes that the collectivization of provision be organized neither in a communitarian nor in an administrative manner, but in the form of a *market,* albeit with two provisos. First, that the exchange of services should take place not through the medium of money, but through *service vouchers* valid only among members and only for the purpose of trade in services between a locally delimited number of households. Secondly, that the coming into being and maintenance of a market of this type, with nonconvertible currency, should be *publicly subsidized,* not financially, but through the provision of rooms, equipment, payments in kind, and human capital. (Offe, 1996: 142)

In collaboration with Rolf Heinze, Offe has engaged in a comparative analysis of the kind of embryonic cooperation circles which are currently to be found. Perhaps the most famous of what Offe and Heinze (1992) call a 'local moneyless exchange system' is LETS. Research about LETS is gradually accumulating as Chapter 9 makes clear, but the point about LETS to underline here is that they are said to have benefits not only for the poor and the unemployed but also for those who wish to pursue alternative, Green lifestyles.

Offe would therefore seem to be suggesting that the policy package which will carry us forward into a post-workfare state and a socially just future has to include both BI and cooperation circles of one form or another. A BI on its own may or may not be efficacious, but one allied to the expansion of non-money exchange systems might form a key element in the third sector of informal economies. Each would strengthen the other: those who wished not to engage in waged labour would not have to rely upon their BI because they would have the opportunity to trade goods and exchange services with others in the third sector; those who routinely engaged in moneyless exchange would also have their BI to fall back upon should their circumstances alter.

Conclusion

BI is often regarded by its supporters as a proposal whose introduction is long overdue and which would have been introduced already were it not for the myopic attachment of the traditional political parties to a society based upon GDP growth and paid employment. To my mind, though, this rather impatient assessment derives from a form of administrative determinism were we can expect BI to emerge once it is widely accepted that both free market capitalism and state welfare capitalism (and all 'Third Way' attempts to converge the two) are failures. So far as Green social policies are concerned this determinism interacts with another: an ecological determinism whereby the increasing evidence of environmental collapse will inevitably drive electorates in the direction of ecological parties, groups and movements. Unfortunately, Green social policies are not going to develop if their supporters sit back and expect the world to fall into their laps. Instead, a combination of short-term pragmatism and long-term idealism is called for, a combination that the Green movement has so far struggled to achieve (hence the familiar distinctions between 'environmentalists' and 'ecologists', 'light Green' and 'dark Green', etc. etc.). This book is premised upon the notion that the subject of social policy might offer an opportunity for more creative alignments and associations to develop.

So far as BI is concerned this means looking not only at the advantages and disadvantages for Greens of a BI *per se*, but also at how BI interacts and overlaps with a variety of other proposals and ideological commitments. Many of these are dealt with throughout this book (and also in Fitzpatrick, 1999a) and if we are not yet in the position were we are able to write a manifesto of Green social policies we can begin to see how and why BI sits at the axis of what such a manifesto may eventually resemble. As such, BI is not only a proposal, it is a site of debate and a space within which the creative combinations that I have just advocated could begin to emerge.

9
The Social Economy and LETS

Colin C. Williams

Introduction

What is the role of the social economy in reforming the welfare state? More particularly, what is its role in tackling the problem of social exclusion and promoting social welfare? The premise of this chapter is that the answers given to these questions differ radically according to whether a mainstream or Green perspective is adopted. As this book has highlighted, mainstream approaches to social welfare systems have tended to adopt a narrow definition of well-being and the means to its achievement. This mainstream view eclipses non-employment forms of work and focuses upon inserting people into formal employment as the principal means of improving their well-being. Assessments of actual and potential welfare policies are then made in terms of their ability or inability to create, or facilitate the creation of, jobs in the formal economy. The starting point of this chapter, however, is that there has been a long tradition in the Green movement of recognising that work is more than employment and that 'means of livelihood' cannot be reduced to earning a living through a job. Instead, the widespread belief in Green political thought is that the development of non-employment forms of work enable more sustainable means of livelihood to be pursued.

Viewed in this light, initiatives in the social economy play very different roles depending upon whether a mainstream or Green perspective is adopted. The argument of this chapter is that although many policy-makers and academics recognise the value of the social economy as a resource for welfare reform (Archibugi and Koenig-Archibugi, 1995; Community Development Foundation, 1995; Fordham, 1995; European Commission, 1996, 1997, 1998; Macfarlane,

1996; OECD, 1996), the rationale which is usually applied is that the social economy might compensate for the deficits of the private and public sectors (ECOTEC, 1998) by providing an alternative source of job creation. The aim of this chapter, however, is to show that adopting such a narrow view of the social economy prevents its wider benefits from being revealed.

To show this, the focus is upon one particularly prominent social economy initiative that is currently receiving much attention in western advanced economies: LETS. Drawing upon recent empirical evidence gathered from a postal survey of all LETS co-ordinators in the UK in 1999, a national survey of LETS members and the results of in-depth, action-orientated ethnographic research conducted over a six month period during 1998-99, in a large LETS in the west of England, this chapter argues that if this initiative is treated purely as a means of generating jobs, then it is likely to have a relatively limited impact. However, if it is regarded in Green terms then it possesses considerably more potential. The implication, therefore, is that policy responses need to move beyond narrow criteria such as job creation in order to recognise and value the full range of ways in which initiatives in the social economy mitigate social exclusion.

Before commencing this evaluation of LETS, however, it is first necessary to identify the nature of the social economy and, following this, to outline the contrasting conceptualisations of social exclusion.

Identifying the social economy

The growth of interest in the social economy has brought to the fore the problem of how to define it. Because its economic activities exist neither in the profit-maximising parts of the market economy nor in the redistributive realm of the public sector, the conventional public-private and state-market dualisms that dominate thinking on how to classify economic activities are no longer sufficient. Instead, terms such as the 'third sector' have become increasingly popular to describe this realm of economic activity (Chanan, 1999). How, therefore, can this social economy or third sector be defined?

One potentially useful approach is to appreciate the historical context of these three sectors and how their roles and functions have fluctuated over time. As Polanyi (1944) asserts, until the collapse of feudalism in Western Europe, exchange between people was formed mainly in accordance with two principles: reciprocity and redistribution. He remarks that it was only after this that the market began to play any real role. But the rise of the market in England during the early 19th century led to strong reactions among various social groups. Workers formed trade unions so as to restrict the free supply of labour and firms formed cartels or trusts to restrict production. Thus, the free market had scarcely come into existence before attempts were made to control it. One consequence was the expansion of the public sector during the 20th century. Even during the advent of industrialism, therefore, the newly emergent 'market' was dependent on the redistributive public sector as regulator and problem-solver. Both systems functioned side-by-side.

Nor, moreover, did reciprocity disappear with the rise of the market and the expansion of the public sector. Although it is often assumed that the principal site where reciprocity occurs, the family, lost its productive functions with the advent of the market, there is little evidence that this was, or indeed has been, the case. Reciprocity has persisted in realms that the public or private sector has not reached, e.g., caring activities. In these realms, such exchange has complemented the private and public sectors. Indeed, the family and kinship network (and, to a lesser extent, neighbourhood and community networks) can still be identified as the principal carrier of the reciprocity principle (Williams and Windebank, 1999, 2000).

However, there is also a quasi-formal realm that this principle of reciprocity inhabits. This is the 'social economy' which addresses those needs and desires that neither the private nor public sectors, nor the informal networks of the family, kin, neighbourhood and community, have managed to fulfil. This social economy possesses four characteristics that distinguish it from the formal private and public spheres (Pestoff, 1996; Lorendahl, 1997; Westerdahl and Westlund, 1998), plus an additional characteristic that distinguishes it from more informal kinship, neighbourhood and community networks:

- It is based on co-operative or closely-related mutual principles;
- It is based on not-for-profit principles in the sense that the initiative does not seek to expropriate a profit from its operations;
- It is private (non-public) in nature even if there is sometimes public sector involvement;
- The tasks conducted by such initiatives include collective economic activities that seek to fulfil people's needs and wants through the production and/or distribution of goods and services;
- Relative to associations between kin, neighbours and friends, it is a formal association that provides an organisational framework for the pursuit of collective self-help activities.

Social economy initiatives, therefore, are *private, formal associations for pursuing economically orientated collective self-help based on not-for-profit and co-operative principles*. Given this definition, we can now examine the various views of its role in combating social exclusion.

The role of the social economy in welfare reform

As we have already indicated, there are two predominant perspectives regarding the role of the social economy *vis-à-vis* welfare reform (Macfarlane, 1996; OECD, 1996; Jordan, 1998; Robinson, 1998; Chanan, 1999; Williams and Windebank, 1999, 2000). Here, each approach is examined in terms of the way it defines social exclusion, the role that the social economy is seen to play and the criteria used to evaluate the potential of social economy initiatives.

The social economy as a springboard into employment

This approach towards welfare reform equates social exclusion primarily with unemployment and social inclusion principally with insertion into employment (Jordan, 1998; Robinson, 1998; Wilson, 1998). To see this, one needs look no further than the numerous work and welfare policies introduced into the UK since the election of the New Labour government in 1997. These include: restructuring the benefits system, e.g. through the Working Families Tax Credit; over-hauling the tax system by, for example, introducing a 10 pence starting rate of tax; 'modernising' national insurance contributions; introducing a minimum wage; and pursuing 'welfare-to-work' policies such as the New Deals for young people, lone parents, the long-term unemployed and disabled (HM Treasury, 1997, 1998; Bennett and Walker, 1998; Hills, 1998; Oppenheim, 1998; Gregg *et al.*, 1999; Powell, 1999). All of these measures are designed to encourage the take-up of paid employment as a solution to social exclusion. Social exclusion is thus approached in a very limited manner seen predominantly in terms of exclusion from employment.

Therefore, the key issue for the social economy in this respect is whether it is capable of providing a new means of employment creation to complement the efforts of the public and private sectors. Indeed, the idea that the social economy can create jobs has steadily gained momentum throughout Europe and North America (Mayer and Katz, 1985; European Commission, 1996, 1998). This attachment of social economy initiatives to the goal of job generation is symbolised by the European Commission naming their new generation of projects the 'Third System and Employment' (ECOTEC, 1998; Haughton, 1998; Westerdahl and Westlund, 1998). Initiatives in the social economy are therefore evaluated according to the following criteria: the number of formal jobs created by such an initiative; its ability to facilitate skills acquisition and maintenance; whether it provides a test-bed for potential new businesses; its ability to develop self-esteem; its ability to maintain the employment-ethic.

The social economy as a complementary means of livelihood

The alternative approach towards welfare reform interprets social exclusion more broadly as the exclusion of citizens from work (both formal *and* informal) and from the income required to meet their basic material needs and creative desires (Williams and Windebank, 1999). While this approach accepts that joblessness is one important way in which people are socially excluded, it views social exclusion as a process, or a set of social relations, between those excluded and the rest of society (Room, 1995; Alcock, 1997; Wilson, 1998). As de Foucauld states:

> It is the denial or absence of social contact which fundamentally distinguishes exclusion. The dignity of the individual derives from integration in a social network–or more precisely, into a system of exchange. An individual brings something to an exchange for the other person, acquires a kind of right, recovers his [*sic*] status as an equal. Social exchange is what provides both a social context and autonomy which are the two essential elements of the individual. (cited in Robbins, 1994: 8)

Individuals, in other words, live in a social world based on reciprocity (Offer, 1997) but reciprocity is not only expressed in the workplace but also in social spheres that nurture reciprocity far away from the world of paid employment.

Here, therefore, the social economy is viewed from the perspective of the informal rather than formal sphere and its role is seen to be to provide access not only to employment but, equally, to other forms of reciprocity that might provide a means of livelihood. Social economy initiatives are thus seen as a means of stemming the degradation of the social fabric in terms of their capability for reciprocal exchange (Macfarlane, 1996; OECD, 1996; Chanan, 1999). This is an important point to emphasise because a number of studies show that reciprocal exchange is damaged (and therefore social exclusion created) due to the ways in which inequalities in the formal and informal sectors reinforce one another. Those excluded from employment also find themselves relatively excluded from using the informal sphere as a coping mechanism (Williams and Windebank, 1999, 2000; Windebank and Williams, 1995). These barriers are as follows.

Economic capital

People often lack the money to acquire the goods and resources necessary to participate in both formal and informal reciprocal exchange. For example, if one cannot afford to socialise there is less likelihood of participating in reciprocal exchanges.

Social network capital

A second barrier to participation in the giving and receiving of aid is that many households have few people that they know well enough to either ask or be asked to do something. This is particularly the case with no-earner households. These households have fewer people to call upon for help either because of their inability to repay a favour, due to their physical inability to do so, i.e., many no-earner households include the physically disabled or retired, or because of the smaller size of social networks that results from being unemployed (Renooy, 1990; Thomas, 1992; Morris, 1994; Kempson, 1996). Given that the long-term unemployed, moreover, mix mostly with other long-term unemployed, have relatively few friends or acquaintances who are employed and are less likely to have kin living locally, the result is that the unemployed have fewer people to call upon for aid than the employed.

Human capital

Many assert that they would engage in more reciprocal exchange if they had more skills and/or different skills, greater confidence, or were physically able. For example, having a formal job often means that customers recognise a person as having a wide range of talents and skills to offer. The level of trust in the ability of a person who is without a job tends to be reduced.

Institutional barriers

A fourth barrier to participation in reciprocal exchange is that many might feel inhibited for fear of being reported to the authorities, whether or not the exchange is paid. Given that working while claiming benefit is normally seen as a more serious offence than tax fraud in contemporary UK society (Jordan *et al.*, 1992; Dean and

Melrose, 1996; Cook, 1997), the result is that this institutional barrier is more likely to act as a constraint on the unemployed, who fear being reported to the social security authorities, than for those in formal jobs. Indeed, the unemployed are often so fearful of being reported for engaging in such work and losing their benefit that many even express a fear of engaging in unpaid exchange in case it is misconstrued (Williams and Windebank, 1999).

Given that the poor and unemployed lack access not only to employment but also informal reciprocal exchange, initiatives in the social economy could be thought of as complementing formal job creation policies. They provide a way of tackling the barriers to participation in all forms of reciprocal exchange experienced by the poor and unemployed. As such, these initiatives are viewed as 'complementary social inclusion policies' that provide an additional means of livelihood based on informal reciprocal exchange and collective self-help.

To evaluate the potential of the social economy, therefore, a wider range of indicators must be utilised. Besides those already listed above (job creation, skills acquisition, etc.), indicators are also brought to bear which evaluate its ability to provide access to types of social inclusion beyond the formal labour market. These all revolve around whether social economy initiatives create new means by which people can engage in reciprocal exchange and develop new means of livelihood outside of the social relations of employment. Here, therefore, we evaluate the effectiveness of a prominent social economy initiative, namely LETS, as a tool for combating social exclusion.

Do LETS tackle social exclusion?

A LETS is created where a group of people form an association and create a local unit of exchange. Members then list their offers of, and requests for, goods and services in a directory and exchange them priced in a local unit of currency. Individuals decide what they want to trade, who they want to trade with and how much trade they wish to engage in. The price is agreed between the buyer and seller. The association keeps a record of the transactions by means of a system of cheques written in the local LETS units. Every time a transaction is made, these cheques are sent to the treasurer who works in a similar manner to a bank sending out regular statements of account to its members. No actual cash is issued since all transactions are by cheque and no interest is charged or paid. The level of LETS units exchanged is thus entirely dependent upon the extent of trading undertaken.

Neither does one need to earn money before one can spend it. Credit is freely available and interest-free.

As such, LETS are very much a social economy initiative. They are private, formal associations for pursuing economically-orientated collective self-help, based on not-for-profit and co-operative principles. They operate in order to fill the voids existing in the provision of needs and wants that are filled neither by the private nor public sectors, nor by the informal networks of the family, kin, neighbourhood and community. These initiatives are currently one of the most prominent social economy initiatives being advocated in the UK. Whichever government document is examined, LETS are often the principal social economy initiative heralded as a potential solution to social exclusion. In the UK, for instance, hardly a government report on social exclusion passes without some mention of these local currency schemes (DETR, 1998; Department of Social Security, DSS, 1998, 1999; Home Office, 1999; Social Exclusion Unit, 1998).

Indeed, by 1999, LETS had spread to many of most advanced economies. There were approximately 303 LETS in the UK, 300 in France, 250 in Australia, 110 in the USA, 100 in Italy, 90 in Holland, 90 in Germany, 57 in New Zealand, 29 in Belgium, 27 in Canada, 19 in Austria, 1 in Switzerland, 14 in Sweden, 7 in Norway and 3 in Denmark, to name but a few of the relevant nations. Up until now, however, there has been no comprehensive attempt to evaluate their effectiveness as a means of tackling social exclusion or to offer lessons on how this could be further improved. Instead, only one-off studies of individual LETS have been conducted (Williams, 1996a, 1996b, 1996c; Pacione, 1997a, 7b; North, 1998, 1999; Seyfang, 1998). Given the frequently contrasting methods used, not only were these studies often not comparable but one did not know whether the findings were unique to the individual LETS or of wider significance.

To evaluate the effectiveness of LETS at both assisting members into formal jobs and providing complementary means of livelihood, three types of research were conducted during 1998 and 1999. First, a survey of all LETS co-ordinators was undertaken in 1999. These co-ordinators were identified from existing databases and, additionally, by snowballing methods. Of the 303 LETS identified and surveyed, 113 responded (a response rate of 37 per cent). Secondly, a membership survey was conducted. From the results of the co-ordinators survey, maximum variation sampling was used to identify widely different types of LETS in existence, e.g. by membership size, urban/rural location, size of area covered, type of members, time established. Surveying

26 LETS, 2515 questionnaires were sent out and 810 (a response rate of 34 per cent) returned. Finally, in-depth, action-orientated ethnographic research was conducted on two LETS in very different locations: the semi-rural area of Stroud and the deprived urban area of Brixton in London.[1] Here, we report the results of these surveys along with the qualitative data from Stroud.

The co-ordinators survey found that those LETS who responded had an average membership of 71.5 members and an average turnover equivalent to £4,664. If these respondents are taken as representative of all LETS in the UK, then the total membership of all UK LETS is approximately 22,000 people and the total turnover equivalent to some £1.4 million. In terms of the total exchange-value of LETS, therefore, these schemes are relatively insignificant. However, when measured in terms of their use-value in tackling social exclusion, as we shall see below, they become more effective vehicles.

Who, therefore, joins LETS and why? Of the 810 members responding to the survey, Table 9.1 reveals that LETS members are predominantly aged 30–49, women, relatively low income groups and those who are either not employed at all or are self-employed. Hence, it appears that the membership profile of LETS is skewed towards the socially excluded if one accepts that non-employment and low incomes are (surrogate) indicators of social exclusion.

Why, therefore, do these people join LETS? Just 2.4 per cent join explicitly to use it as a means of gaining access to employment and these are all people who are, or seek to become self-employed. The rest join either for ideological reasons (23.3 per cent) or to engage in 'complementary' means of livelihood for both economic and community-building reasons (74.3 per cent). Economic reasons are cited predominantly by the relatively poor and unemployed, who articulate demand-side rationales such as their lack of money to buy goods and services in the formal economy (12.0 per cent of members), their desire for a specific service (9.3 per cent) or their wish to exchange goods and services (20.8 per cent). Only 1.3 per cent of members state a supply-side rationale and, when they did, this was always that they wished to use their skills. Meanwhile, 22.9 per cent of all members, who are usually the employed and relatively affluent, cite community-building rationales.

Given this concentration of low-income households and non-employed people in the membership, and the fact that they joined predominantly to develop complementary means of livelihood for economic (rather than community-building) reasons, we now evaluate

Table 9.1   Characteristics of UK LETS members

	%
Age group:	
Under 20	0.6
20–29	5.9
30–39	25.1
40–49	29.6
50–59	22.4
60–69	11.5
70 or over	4.9
Gross Annual Household Income:	
<£4,160	8.4
£4,161–6,499	8.2
£6,500–9,099	12.3
£9,100–4,299	22.3
£14,300–9,299	14.4
£19,500–24,699	11.2
£24,700–33,799	12.9
>33,800	10.4
Gender:	
Men	31.3
Women	68.7
Employment Status:	
Full-time employee	19.5
Part-time employee	18.2
Self-employed with employees	2.7
Self-employed without employees	25.3
Voluntary worker	1.5
Registered unemployed claiming benefit	4.6
Home-maker	7.7
Student	1.5
Retired	14.3
Permanently sick	3.7
Other	0.7

the effectiveness of LETS, first as springboards into employment and then as complementary means of livelihood.

Evaluating LETS as a springboard into employment

To evaluate LETS in this respect, one must examine the number of formal jobs created by LETS, whether directly or indirectly, their ability to facilitate the acquisition and maintenance of skills, whether they provide a test-bed for potential new businesses, their ability to develop self-esteem and to nurture the employment ethic (Fitzpatrick, 1998).

Since LETS are run by volunteers they are directly responsible for the creation of few jobs. Nevertheless, some 4.9 per cent of members surveyed asserted that LETS had helped them gain formal employment. This was entirely because working in the LETS office administering the scheme had enabled valuable administrative skills to be acquired which they had then been able to use to successfully apply for formal jobs. In no other way, however, did the LETS enable participants to use it as a springboard to become an employee. As such, the significance of LETS in this respect is limited since only a small number of people can at any one time play a prominent role in administering the scheme.

For others, however, LETS represented a direct or indirect springboard into *self*-employment. Some 10.7 per cent of respondents asserted that LETS had helped them become self-employed by enabling them to develop their client base (cited by 41.2 per cent of those who were self-employed), ease the cash-flow of their business (cited by 28.6 per cent) and enabling them to use it as a test-bed for their products and services (cited by nearly all who defined themselves as self-employed). As several respondents stated during the interviews and/or focus group discussions:

I was looking to start off as a freelance journalist, at the time, and it [joining LETS] was just another way of generating some work and some contacts and building up experience without having to put in, sort of, the risk of hard currency. (man aged 35–39 in focus group discussion)

I became a LETS member and used the LETS as a source to advertise my services and from this I have managed to go self-employed. All of my customers are coming through the LETS and my business is slowly building up. The LETS has been extremely important in this development both financially and the community support it provides – I get my childcare paid for through the LETS which enables my business development, LETS has enabled my survival. At the moment life is very tight, I'd be desperate without LETS. (interview with self-employed single parent aged 35–39 on Income Support who set herself up as a self-employed massage therapist and had transferred to Family Credit)

More indirectly, some 24.8 per cent of all respondents asserted that the LETS had boosted their self-confidence. This was particularly the case among younger people, the registered unemployed and those with few qualifications. In addition, 15.9 per cent stated that it had enabled new skills to be acquired (24.3 per cent of the registered unemployed), mostly relating to computing, administration and interpersonal skills. By recognising and valuing work beyond employment, therefore, LETS was not only helping members gain entry to employment or develop self-employed business ventures, but also providing them with greater personal transferable skills and self-confidence which would later be of use to them in the formal labour market. As a 50–54-year-old unemployed single woman put it:

> Coming into LETS I've had a lot of interaction with other people, lots of different people, and it helps me with my confidence. I'm going to learn how to do the directory, and I've been inputting cheques into the computer accounts so I'm learning different things through my LETS work. I think I just enjoy the contact with other people and the fact that I'm getting LETS responsibilities now, it makes me feel that I'm a bit important and getting invited to meetings, it's really good. And writing up messages in the day book, someone put 'good idea, well done' – well it just makes you feel valued and that you are making a contribution. …. I've been out of work for over two years and I've had problems getting references from previous employers because they say that can't remember that long ago, which is upsetting … so I should be able to get references from the LETS for the work I'm doing, which will help in looking for paid work when I'm ready.

LETS, in consequence, do appear to be a useful springboard into the formal labour market for a small but significant proportion of members. Within the logic of this employment-based perspective, therefore, several possible policy responses are suggested by this research. If LETS administration was to be recognised and subsidised by the 'voluntary and community' aspect of New Deal, this would provide workers with a proven means of entering the formal labour market as employees and, at the same time, enable the more efficient running of the LETS (since it would not be so reliant on 'volunteers' for its day-to-day administration). Second, it appears that many currently operating as self-employed in LETS could be both encouraged to enter the 'self-employment' option in the New Deal and their trading on LETS could be recognised as part of their attempt to become self-employed. As we shall see later, however, such policy responses would be strongly resisted by many LETS members who contest the interpretation of the social economy that underpins this perspective as well as its potential impacts on LETS.

Evaluating LETS as a complementary means of livelihood

In order to evaluate LETS as tools for facilitating complementary means of livelihood, the extent to which they counter the barriers discussed above (economic capital, social network capital, etc.) need to be analysed.

Starting with the extent to which LETS tackle the barrier of economic capital, some 40 per cent of overall members assert that LETS provided them with access to interest-free credit (though 62.1 per cent of the registered unemployed and 51 per cent of low-income households asserted this). LETS, therefore, provide people with access to money. For two-thirds (64.5 per cent) of the registered unemployed, this had helped them cope with unemployment, with some 3.1 per cent of their total income coming from their LETS activity.

LETS also enable participants to tackle the barrier of social network capital. Some 76.2 per cent of respondents asserted that the LETS had helped them to develop a network of people upon whom they could call for help while 55.6 per cent asserted that it had helped them develop a *wider* network of friends and, for 31.2 per cent, *deeper* friendships. LETS, therefore, develop 'bridges', i.e. bringing people together who did not know each other before, more than 'bonds', i.e., bringing people who already know each other closer together. Given that most members lacked kinship networks in the localities they inhabited, and that kinship networks are the principal source of mutual aid in contemporary society (Williams and Windebank, 1999), LETS thus provide those without such a local network with a substitute. Some 95.3 per cent of LETS members, that is, had no grandparents living in the area, 79.5 per cent no parents, 84.3 per cent no brothers or sisters, 58.2 per cent no children, 92.6 per cent no uncles or aunts and 90.8 per cent no cousins.

This important role that LETS play in developing social networks was brought out in numerous interviews and focus group discussions. Take, for example, the following extract from a focus group discussion:

Discussant 1:
We joined LETS only a year ago, but moving to a new area you don't have your family and friends readily laid on, and it's a very good way to get to know people. ... If you haven't got anybody or you don't know your neighbours very well then it's a great way of asking people to do something a bit silly that you wouldn't be able to do. The first person we contacted, we wanted something moving and we couldn't lift it ourselves and we thought oh we've got no

neighbours, or they're old neighbours, so it was sort of the introduction to LETS (part-time employed woman aged 30–34).

Discussant 2:
We didn't actually know anyone to help us carry something (full-time employed woman aged 30–34).

Discussant 1:
So it was as simple as that, so that was the starting point and now we're just looking to get into debt and spend more LETS and get involved that way.

As such, LETS can be a way of creating a form of 'bridging' social capital. However, for some people, especially unemployed members, who would otherwise have relatively few opportunities to forge new social networks, it is also being used to develop 'bonding' social capital. As an unemployed single woman aged 50–54 stated during an interview:

When I first moved here, I was finding it very, very hard to meet people, make friends, people are very reserved around this area, they don't sort of welcome strangers with open arms; so I thought I could meet people through the LETS system, and that's worked really well … you see that's the one thing about being on benefits, low-income, it's (LETS) exceedingly useful, it's a way of instead of barely existing, you know it enables you to do a lot of things, and its good, and like I was very isolated when I first moved here and through the LETS I don't feel so isolated at all now, I've got lots of people I know to speak to, and I've got a couple of very good friends, it's great, there's a network of people available.

Besides tackling the barriers of economic capital and social network capital, there is also evidence that LETS tackles the barrier of human capital that can constrain participation in reciprocal relationships. As discussed above, LETS provide an opportunity for people to both maintain and develop their skills as well as to rebuild their self-confidence and self-esteem by engaging in meaningful and productive activity that is valued and recognised by others who display a willingness to pay for such endeavour.

Finally, there is an institutional barrier to pursuing complementary means of livelihood. Many who are unemployed are fearful of being reported to the authorities, even if they engage in unpaid mutual aid.

This is not currently being overcome by LETS. Although, overall, only 13 per cent of members feel worried about tax liabilities and 12 per cent about reductions in welfare payments, 65 per cent of registered unemployed members are concerned about their situation. Similarly, and as identified in a survey of 103 non-members in Stroud, grave concerns exist amongst unemployed non-members that their trades will result in a reduction in benefits and this makes them wary of joining. Ironically, therefore, those who would most benefit from LETS are discouraged from joining and trading due to the uncertainty over their legal position *vis-à-vis* the benefits disregard. This is something that, so far, New Labour have failed to take into account (Home Office, 1999).

Conclusions and policy implications

In order to evaluate LETS as a vehicle for tackling social exclusion they have been analysed here in the context of two contrasting conceptualisations of social exclusion. The first equates social exclusion with unemployment and regards the social economy as a potential springboard into employment for the unemployed; the second interprets social exclusion more broadly as the exclusion of citizens from all forms of reciprocal exchange, and so from the work and income required to meet their basic material needs and creative desires, and sees the social economy much more as a 'complementary social inclusion policy' which provides people with access to a means of livelihood beyond the formal economy.

The foregoing analysis revealed that LETS has significance for a wide proportion of members as a vehicle for providing complementary means of livelihood and for some it provided both this *and* acted as a springboard into employment. Consequently, if social exclusion is equated solely with unemployment and the social economy is viewed merely as an addendum to the formal economy, then the full value of such initiatives will not be recognised. However, if social exclusion is more broadly conceptualised as the exclusion of citizens from work (both formal and informal) and income to meet their basic material needs and creative desires, then the broader significance of such social economy initiatives can be realised.

The current problem, though, is that policies towards initiatives such as LETS are predicated on the narrower of the above premises. Grounded in an employment ethic that views formal jobs as the only means of social inclusion, rather than a 'work' ethic that recognises other avenues beyond employment by which people can seek inclusion (Fitzpatrick, 1998), the whole thrust of the current policy agenda on LETS in particular, and social economy initiatives more broadly, is based on their potential for job creation and preparation. Of course, even within the limitations of this perspective there is much that can be achieved. As we saw above, LETS do have some positive implications for job-creation and, in particular, self-employment. These are aspects of LETS that have so far remained unexploited by UK governments.

However, it has to be remembered that such policy responses would be strongly resisted by many LETS members. Most members, as shown above, join LETS to forge complementary means of livelihood. The qualitative aspect of the project, moreover, suggests that tying LETS into welfare-to-work measures is perhaps neither the most appropriate use of resources nor necessarily what current LETS members would desire. When funding was raised as an issue by members, the desire was for a LETS development worker to be resourced to do 'outreach' or community development work, not the basic office/administrative tasks, in order to build bridges between the LETS and voluntary and community sector groups in the locality, and to widen participation in the scheme. Such a funded worker, it was felt, could help facilitate specific projects, e.g. a LETS shop on the High Street run by members for the sale of their goods and services, in order to make LETS more visible within the local community. Any policy, therefore, that merely resourced placements for office workers (which necessarily would have a limited throughput given that there are merely 303 LETS in the UK) and facilitated individuals to work under the self-employment option of the New Deal would not only be seen by LETS members as failing to further facilitate its strategic desires to promote social inclusion but would also be seen as diminishing the sense of collective ownership which prevails in the LETS due to the voluntary nature of its operations. Before pursuing solutions which treat LETS merely as a springboard into employment, therefore, greater consideration is needed of the (potentially damaging) impacts of such policies upon the dynamics of LETS.

In the final analysis, it might be that LETS should be assessed both in terms of job-related criteria (the mainstream view) and in terms of criteria derived from the notion of the social economy as a 'complementary alternative' to the formal economy (the Green view). In sum, for the full range of ways in which social economy initiatives mitigate social exclusion to be recognised and valued it must be accepted that social inclusion does not solely mean insertion into employment and that a critical re-evaluation of the meaning and scope of work and welfare in contemporary society is required. Indeed, at the start of the new millennium, it is perhaps only by casting off the hegemonic totalising discourse of employment as the only route out of poverty, and thinking forward to how we want work and welfare structured in this new century, that new forms of coping mechanism for the socially excluded can be designed and implemented. Indeed, there is much to be learnt from LETS members themselves in this respect who are using this social economy initiative as a vehicle for providing themselves with complementary means of social inclusion beyond employment, rather than merely as a vehicle for helping them into formal employment.

Acknowledgements

The author would like to thank the Economic and Social Research Council (ESRC) for funding this research project (R000237208) as well as all of the national LETS co-ordinators and members who responded to the surveys and the membership of Stroud LETS for being so open during the six-month visit of the researcher.

10
Working Time Reductions

Adrian Little

Introduction: work in the new capitalism

The role of work in advanced capitalist societies has become increasingly important in recent years as we witness the advent of new forms of political economy. Although Western economies have become more 'post-industrial' (Little, 1998) this does not entail the deterministic creation of post-industrial *society* (Bell, 1973). While Bell's thesis may be partially vindicated by the increasing tertiarisation of Western economies, his prediction of the declining importance of work within wider society has not taken place. Indeed, it is arguably the case that the decline of orthodox perspectives in political economy on work have taken place alongside a process whereby it has become an even more important denominator in understanding contemporary society. In this context Greens have been at the forefront of debates on the significance of paid work and its contribution to social welfare.

The continued centrality of work in contemporary Western societies has been exemplified by a variety of studies that have recognised the problems associated with the declining availability of work and the impact that this process has on social integration and individual well-being (Rifkin, 1995; Sennett, 1998; Gorz, 1999; Beck, 2000). Sennett examines the ways in which forms of work associated with the rise of the new, flexible economy undermine traditional sources of social insertion and identity in capitalist societies. Thus, 'short-term capitalism threatens to corrode ... character, particularly those qualities of character which bind human beings to one another and furnishes each with a sense of sustainable self' (Sennett, 1998: 27). From this perspective work can perform an integrative function by enabling individuals to contribute to wider society and acquire virtues which lead to social identity and responsibility. For Sennett, this potentially positive dimension of work is gradually being diminished through the growth of flexibility in the economy and the attendant short-termism that comes to dominate social and economic organisation.

In the work of Rifkin (1995) and Gorz (1989, 1994, 1999) there is a similar understanding of the impact that these economic developments can have on social integration and individual character. However there is also recognition that changing forms and patterns of work also offer potential opportunities for altering the ways in which we understand social cohesion and identity formation. Thus, for example, in Gorz's theory there is a strong emphasis upon the social and economic benefits that could accrue if political initiatives were employed to take advantage of new technology. In this vein he argues that we could use the changing technological processes in the new economy to emancipate people from the work ethic. According to Gorz, then, technological developments create spaces for a redefinition of the centrality of paid and the importance of unpaid work and leisure. Moreover, if harnessed in a progressive fashion, opportunities exist to rethink the respective roles of markets and planning with the distinct possibility of realising the objective of establishing a genuinely socialised economy.

These analyses of the changing face of work in capitalist societies imply that economic changes do not determine social relations. Rather the latter are dependent upon political decisions that are made regarding the way in which technological developments impact upon wider society. Indeed, in the work of commentators such as Gorz there is the suggestion that, should the appropriate political will exist, technology could be employed to reduce working hours for all. The potential benefits of such a policy are manifold but they include the possibility for individuals to acquire a self-determined balance of paid and unpaid work on the one hand and the possibility of reducing unemployment by redistributing work on the other. These ideas will be analysed in the course of this chapter through discussion of Green and socialist agendas on work and the desirability or otherwise of policies for working time reductions. In so doing we will examine the mechanics of a policy of reducing working hours and evaluate the example of France under the government of Lionel Jospin where such a policy was introduced. From this, the utility of working time reductions for Green political discourses will be discussed in the light of the contribution such a policy could make to social welfare.

Arguments for working time reductions

The idea of working time reductions has a chequered history in the traditions of socialism and political ecology. Although the concept of reducing the centrality of paid work and the work ethic can be implicitly traced to Marx, early anarchists and utopian socialists in the 19th century, there is an equally strong tradition on the left of what can be regarded as an economistic focus on work. This can be identified in orthodox Marxist perspectives that hinge upon the labour power of the proletariat as the major weapon of the disenfranchised against the system which exploits them. In social democratic thinking, in the post-1945, era this economism took the form of the commitment to full employment (although this was primarily regarded as full male employment) whereby the touchstone of social democracy became the provision of sufficient full-time employment to allow the overwhelming majority of men to work and earn a family wage. Both of these perspectives retain considerable influence in leftist thinking – orthodox Marxists have never lost their focus upon the labour power of the proletariat whereas social democrats such as British Chancellor Gordon Brown rediscovered the idea of full employment in the aftermath of the Thatcherite backlash. In its current guise, social democracy has advocated a New Keynesian approach whereby the emphasis is on full employability and the supply-side of the economy rather than the traditional concern for effective demand.

Against this background there has always been limited space for those on the left who have sought to put forward the idea of working time reductions as a strategy for furthering socialist objectives. This trend has been notable in the treatment of the theoretical work of Andre Gorz (Little, 1996) which has never been accorded the significance it merits. It should be with a degree of irony that we have witnessed the strategy of the Jospin government in France and elements of the trade union movement in Europe (especially Germany) where the ideas of working time reductions have become commonplace – these are strategies that Gorz has consistently advocated for the last 25 years only to be neglected by the left in both its radical and mainstream traditions. For this reason Gorz has had as much influence in forging the direction of environmental thinking on work as he has had on the left, as Greens have been more willing to countenance radical ideas especially those that have a clear future-orientated dimension.

The social democratic concern for full employment has been a traditional feature of left of centre governments and parties especially in the era between 1945 and 1970. Buoyed by post-war reconstruction and the hegemony of Keynesian ideas on economic management, full employment became a touchstone for social democrats. From the late 1960s however the idea came under increased critical focus from a radical left that challenged the very notion of full employment. Simultaneously there was a neo-liberal backlash which suggested that such a goal was unachievable and that, even if full employment could be realised, it would be economically detrimental in the longer term due to the excessive state intervention that it would involve. Whilst opposing the philosophical principles behind neo-liberal theory, many on the left and Greens have taken on board the idea that full employment may not be sustainable in the long term and that alternative strategies are required.

It is ironic in this context that the last decade of the twentieth century saw a rebirth in theoretical advocacy of full employment (Hutton, 1996). However, in its new manifestation the advocacy of full employment varies significantly from older, more traditional versions. This is evident in the variety of policies and ideas (many of them focused on the supply-side of the economy) put forward to increase employability in the current economic climate (Philpott, 1997). These include new understandings of the changing gender balance of paid work, the use of tax credits, the shifting relationship between work and welfare towards 'workfare', and the possibilities that emanate from strategies such as work sharing. Moreover, there is much evidence that important debates on employment and unemployment are taking place across the European Union (Compston, 1997). However, what is not as evident on the social democratic agenda is the possibility that not only do we have opportunities to reinterpret full employment, we also face choices regarding the very role that different types of work should play in a more sustainable economy. Thus social democrats have been somewhat negligent in applying their ideas to the agenda on reduced working hours that has emanated mainly from radical left-libertarians and some political ecologists.

Whereas proposals for working time reductions hold considerable sway in ecological thinking, commentators such as Andrew Dobson (2000a) have noted how Green theory has frequently invoked something of an ascetic work ethic. Thus the Green agenda on the role of work is far from clear cut and there are forms of working-time reduction that do not command support in ecological thought. First, Greens are critical of approaches to working time reductions that suggest that technological change will deterministically manufacture a leisure society. Dobson recognises that, while attitudes towards work vary in Green theory, there is a reluctance to embrace the economic factors that give rise to many of the debates on reduced working hours. This feeds into the second criticism that corresponds with sceptical perspectives on the post-industrial society thesis. Greens argue that the society envisaged by post-industrial theorists, of expanded knowledge, information and leisure, would still be predicated upon industrialist attitudes and the mechanisms of capitalist economies that give rise to so many environmental problems. Thus they see little progressive in the development of the 'post-industrial society thesis' or the workless future because it would still be predicated upon the dominance of paid employment as the major source of social standing and inclusion. In other words, they tend to recognise that the narrow definition of work will continue to hold considerable sway over the organisation of post-industrial economies and social institutions such as the welfare state, regardless of moves towards working time reductions. Dobson notes that this conflicts with some of the basic assumptions about work in ecological thought:

> Greens will be sceptical (at the very least) of the workless future because they think work is a good thing to do. In this respect they are part of a tradition that has it that work is a noble occupation, that it uplifts the spirit and helps create and reproduce ties with one's community – even helps to create oneself. This view has it that work is an obligation both to oneself and one's society and that this obligation has to be redeemed. (Dobson, 2000a: 109)

The dominant position on work within political ecology is one of scepticism about the dominant ideology that work equates with paid employment alone. Thus Dobson (2000a: 92) states that, for Greens, 'their renegotiation of the meaning of work leads them to suggest ways of "freeing" it from what they see as restrictions founded on the modern (and archaic) sense that work is just paid employment'.

In this context, and like many socialists, Greens must grapple with the underpinning message of proposals for reduced working hours and this is the origin of the third criticism they raise. Clearly there are ways in which reduced working hours could be constructed as a strategy to meet some of the realities of modern flexible economies, as was the case in the policy employed by Jospin's government in France. However, there needs to be a different rationale if strategies are going to harness support within political ecology. In this sense, working time reductions that are based upon contemporary economic imperatives and are designed to shore up globalised markets are less likely to find advocates within the Green movement than proposals that are couched in terms of raising the priority given to unpaid work, voluntary activities and work in the local community. Thus Green support for working time reductions has at its centre the idea of reducing the centrality of paid employment in the formal economy and providing greater support for activities that do not necessarily make a contribution to economic wealth. The fact that this may be achievable in the context of shoring up flexible markets and global capitalism is not a particularly strong contribution to the Green argument especially amongst more fundamentalist elements within political ecology.

The Green critique of the primacy of paid employment also feeds into their analysis of the relationship between work and welfare. In the light of modern technological and economic developments, opportunities exist to do more with less. Productivity increases may reduce the amount of working time required to produce goods and the automation of the labour process may reduce the amount of human labour needed in the workplace. With this in mind the continuation of the ideology of paid work, whereby paid employment is accorded a higher status than any other form of activity, seems increasingly anachronistic. This demonstrates the political nature of decisions over how technological changes affect us. The workerist basis of welfare arrangements in most Western industrialised nations (especially the United Kingdom) suggests that governments have an interest in the maintenance of work-based welfare strategies. Thus governments and politicians appear to have a vested interest in the promotion of paid employment, the demonisation of the unemployed, the under-valuation of caring and child-rearing, and the continuation of the distinction between the formal and informal economies (see Chapter 9). Against this perspective Greens tend to argue that

> most social security systems (and certainly Britain's, based on Beveridge's 1942 proposal) have been designed around the assumptions of a growth economy and a system of reward based on the existence of practically universal paid employment. Once those assumptions no longer hold (and Greens believe that they do not), the social security system based upon them must come into question too. (Dobson, 2000a: 111)

In this vein it is important to analyse the kinds of proposals that have been put forward to bring about reductions in working time. This involves an analysis of the impact of technology on the workplace (and thus the economic choices which they afford us) and the potential for a radically different relationship between the way we work and the welfare regimes that are set in place.

Gorz's political economy

As with the notion of paid work itself, there is some ambiguity within the Green tradition about the nature of technological development. Technology is frequently treated with scepticism, primarily because the historical development in technical expertise has been subject to mainly economic imperatives. Too often, the impetus for technological research and development has been seen as a by-product of the pursuit of growth, profit and capital accumulation. In other words, the dynamic behind the use of technology has been driven by economic rationality. However, the actual rationality behind this dynamic is less than clear. The increasing use and sophistication of technology should, in purely economically rational terms, lead to a reduction in the amount of human labour required to produce goods and services. However, the experience of technological development in the industrialised world has been one where human labour has remained fundamentally central to our economic arrangements. Indeed, where technology provides opportunities for us to rethink our relationship with paid work, the pursuit and availability of such work has become increasingly paramount as sources of both social and economic inclusion and participation. As noted above, it is this scenario that Greens and elements of the libertarian-left have challenged as irrational and an obstacle to emancipatory progress away from work-based society. In short, Greens oppose the ways in which technology is made to serve the ideology of paid work but recognise the capacity of technology to serve more progressive ends.

The most prominent commentator in this area has been Andre Gorz who, despite his influence on the Green movement, has clearer left-libertarian credentials. His advocacy of reduced working hours has been a long term project but his most systematic exposition of his proposals is to be found in *Capitalism, Socialism, Ecology* (1994). There, Gorz explains not only the theoretical validity and requirement for working time reductions, but also the economic case for such a policy. He explains a variety of ways in which reduced working hours could be implemented, involving differing strategies for combining the policy with changing levels of productivity and varying proposals for the impact of such changes upon wages. In this sense, Gorz's work makes clear that the idea of reduced working hours is not part of a zero-sum game whereby there are inevitable fundamental knock-on effects on the rest of the economy. Rather, what Gorz's theory makes clear is that working time reductions can only be implemented within a bundle of economic strategies and policies that will impact upon the overall economic and social outcomes. One might add that Gorz, as we shall see, provides a number of differing optional models that are explicitly hypothetical. Rather than criticising them as such, political theorists and economists need to recognise that the impact of hypothetical models of economic change are subject to the political context in which those changes are implemented and experienced. In other words, economic objections to working time reductions must be contextualised within the real conditions in which policies are operationalised. With this in mind, however, it is important to examine the theoretical models that Gorz puts forward to explain the available options that a strategy of reducing working hours could bring to bear.

In a pivotal chapter in *Capitalism, Socialism, Ecology* Gorz explained how his theoretical beliefs translated into a political economy which promised 'Shorter Hours, Same Pay'. The argument constructed is an attempt to counteract the claims made by economic liberals and some within the trade union movement that it was not possible to reduce working hours without a concomitant loss of earnings – for a more detailed discussion of Gorz's position in this area, see Little (1998: ch. 6) and the example of working time reductions without loss of pay in France is examined blow. Gorz rejects the belief, prominent even amongst sympathetic economists and trade unionists, that there would have to be substantially reduced salaries for educated, professional workers. Moreover, he suggests that not only can we reduce working hours but we can also create more jobs without wage reductions being necessary. What we require, according to Gorz, is to free our imaginations from economic orthodoxy. This would allow us to entertain the possibility that 'for as long as not only productivity but also output continue to grow – even slightly – it is possible to do everything at once: reduce working hours, bring about a fall in unemployment and maintain, or even increase, incomes' (Gorz, 1994: 103–4). Gorz provides a number of projections concerning ways in which a planned increase in both production and productivity could be channelled into the maintenance of current working hours, of current employment

Table 10.1   Gorz's strategy for reducing working hours and raising wages

Option	Workforce	Wages	Hours
1. Maintain current working hours	Reduce by 4%	Increase by 12%	No change
2. Maintain current employment levels	No change	Increase by 8%	Reduce by 4%
3. Maintain current wage levels	Increase by 8%	No change	Reduce by 12%
4. Reduce hours, increase wages and workforce	Increase by 5%	Increase by 3%	Reduce by 9%

levels, of wages at current levels, or, his preferred option, the reduction of working hours while increasing wages and the workforce.

To demonstrate his argument Gorz (1994: 104–8) sets out a theoretical example, based on the then standard French 39 hour working week, in which, over the course of four years, production is increased by 8 per cent and productivity by 12 per cent. In this scenario, 108 per cent of current production could be produced with 96 per cent of the quantity of labour (100 + 8 – 12 = 96). For Gorz, this would provide us with a range of options as to how to implement these changes (see Table 10.1).

Gorz recognises that this does not exhaust the available economic options and that it is a hypothetical model. In this sense the real conditions in which strategies for working time reductions can be implemented will depend upon a range of economic factors that will impact upon the particular policy that is most suitable. The point of these projections is to show how our economic arrangements emanate from political choices and that it is a matter of strategy and political will as to how economic possibilities are translated into social reality.

To demonstrate this point Gorz also provides examples of the choices that might result from an ecological restructuring in which there was reduced or zero economic growth. He does not favour the latter but does recognise the economic choices that such a strategy might engender. In the short term, he prefers to promote a system in which continued growth would provide the benefits of shorter working hours and increased pay (option four in Table 10.1). In planning for the longer term, he suggests a foundation would then be in place from which an ecological restructuring could take place 'enabling us to live better while consuming, producing and working less' (Gorz, 1994: 108). Gorz suggests that, should it be necessary to reduce wages due to the implementation of lower working hours, then there should be a 'second cheque' (a term coined by Guy Aznar) to compensate workers. This would be distributed by the state and funded out of taxing consumption rather than income or productivity gains. Gorz suggests that this could be achieved via VAT or by focusing on specific products. Importantly he does not deal with the politically sensitive question of whether the taxing of consumption will hit the poor disproportionately or beyond the capacity of a second cheque to compensate. However he does imply that he envisages the taxation of consumption of products that are potentially harmful as his priority. Here he cites the examples of fuel, motor vehicles, luxuries, alcohol, etc. (Gorz, 1994: 111). The suggestion, then, is that difficult decisions need to be made regarding the specific focus of taxation but that there are some products that Greens are likely to see as particularly appropriate for incurring these duties. As with many such environmental or left-libertarian proposals, the political obstacles to the implementation of these strategies are clear regardless of their perceived necessity. Nonetheless, Gorz is aware of the *realpolitik* of demonstrating the advantages of working time reductions and that is why he argues for the implementation of such measures before we seek stricter limitations on growth. In this sense his proposals are relevant to, rather than a blueprint for, a Green political economy (Little, 2000).

The aim of reducing working hours is usually placed in the context of new forms of welfare strategy, such as those being examined elsewhere in this book. Without reiterating all of these, it is worthwhile examining the ways in which proposals such as a Basic Income might be complementary to a broader economic strategy of reducing working hours (see Chapter 8). Of interest here is Gorz's most recent work in which he adopts a much less sceptical stance towards Basic Income theory than was previously the case in his attempt to theorise ways of moving beyond wage-based society. Thus, in his later work, he sees Basic Income as a key factor in a process in which 'working time will cease to be the dominant social time' (Gorz, 1999: 73. This corresponds with his advocacy of 'multi-activity' whereby individuals are given real opportunities to engage in a range of activities rather than being bound within the constraints of paid labour and the ideology of paid work. His concern for individual autonomy remains the driving force behind his theoretical ideas and he sees Basic Income as part of a recipe that would 'increase as far as possible the spaces and resources ... which allow modes of life, co-operation and activities to emerge that lie outside the power apparatuses of capital and the state' (Gorz, 1999: 79). Implicit in this advocacy of a meaningful BI is a rejection of both the workfare strategies that have been embraced by some social democrats and minimalist guaranteed incomes that appeal to many neo-liberals. Thus Gorz maintains elements of his older critique of Basic Income insofar as he wants to ensure that it does not become a subsidy for ruthless employers or a means of shoring up the flexibility 'demanded' by global markets. In other words, Basic Income should not be a mere subsistence payment but must guarantee real opportunities for autonomous choices over activity.

In his earlier work Gorz argued for a form of guaranteed income but one that was tied to the performance of socially necessary work over the course of a lifetime (Little, 1998). Although this was not to be a strictly reciprocal link, it did contain an element of conditionality, albeit a weak one. Nonetheless, he now argues that the changing nature of work in post-Fordist economies has altered conditions in such a way that the linkage between work and income (which he has always criticised) must be broken fundamentally. In the current climate Gorz defends unconditionality on the grounds that working time is no longer the measure of labour in the new economy and that the only way to maintain work requirements would be to 'economise' and standardise activities such as voluntary work or domestic responsibilities. These are the very activities that he has always wanted to protect from the logic of economic rationality (Gorz, 1989). Thus, he sees an unconditional Basic Income as a means of maintaining the essential non-economic nature of certain social activities. Moreover an unconditional income would also protect individual autonomy on a number of levels in the context of economic change that often leads to a decline in self-determination as work and welfare are currently organised.

This demonstrates that, for Gorz, Basic Income is part of a package of economic and social measures (and only valuable in that context) that includes opportunities to work and not to work and the provision of environmental spaces to facilitate a range of possible actions. Importantly, in environmental terms, Gorz's later position opens up the possibility of reduced levels of economic growth that he regarded as politically undesirable in *Capitalism, Socialism, Ecology*. His argument in *Reclaiming Work* implies that there could be a reduction in working hours with much lower levels of economic growth. In this scenario, hypothetically, we could envisage a reduction in income from paid work because the loss of wages could be offset by the Basic Income.

However, Gorz's original point, that it may be politically difficult to get the wealthy and powerful to buy into this model as they would appear to have most to lose from such a system, remains powerful. Thus he continues to recognise that these strategies can come up against what he calls 'the *problems of the political*' (Gorz, 1999: 110). However, after the election of the Jospin government in France in 1997 we witnessed a government that was prepared to use state mechanisms to organise the economy in such a way as to make working time reductions feasible without a loss of earnings. Whilst the strategy employed in France has not been based on ecological thinking – based as it is on shoring up employment levels in the context of flexibility and global markets – it does demonstrate that political obstacles to working time reductions are not insurmountable.

The French example

Working-time reductions have long been at the forefront of the agenda of the French left and that influence has filtered through into mainstream discussions of the future of work such as that in the 1995 *Boissonnat Report* (Gorz, 1999: 74–5). Not only has this influence been notable in academic literature but the proposal has also been manifest within debates in the Socialist Party and trade unions since the early 1980s, although the promise of a 35 hour week was not realised at that time (Milner and Mouriaux, 1997). The idea of working time reductions and policies such as work-sharing were given a new dynamism by the growth of the French Greens during the 1990s (Laborde, 1999: 164). Nonetheless, despite various initiatives by previous socialist governments, these ideas did not reach any real fruition until the election of the Socialist government (in coalition with the Greens and the French Communists) under Lionel Jospin in 1997. The immediate problem the government faced was the difficult task of tackling high levels of unemployment. On top of this was the associated issue of social division and insecurity, recognised across the political spectrum, and regarded by many on the left as a product of the prevalence of insecure jobs (Girling, 1998: ch. 12).

Unemployment in France was as high as 12.5 per cent in 1997 and was regarded as a major blight on society and the economy (Ardagh, 1999: 175). Early signs suggest that Jospin's strategy has succeeded insofar as unemployment had declined to under 10 per cent by 2000 (Islam, 2000). To attribute this to the working time reductions alone, however, would be erroneous. Islam (2000) notes that the French economy was in 'rude health' with low inflation, a strong stock index and the fastest growth of all major European economies. Similarly, Dumas (2000) has pointed out how French growth was 'driven by domestic demand' rather than exports and that the 'mainspring of the French recovery is jobs'. Of course the creation of more jobs was a success story for the Jospin government but the improvements in the French economy also reflect the rather low base from which recovery began. More startling then is the prediction of Pierre Alain Muet, a key economic adviser to Jospin, that he favoured the implementation of a Negative Income Tax that would contribute to a scenario in which 'I expect France to return to full employment, unemployment below four per cent, in the second half of this decade' (quoted in Islam, 2000). Clearly, then, the French economy is buoyant and on course for further strengthening in the future following the lean years from 1981 to 1997. How, though, have working time reductions contributed to these improvements?

The strategy put forward by Jospin's government was to implement a 35 hour week for large firms by 2000 and for smaller companies with under 20 employees by 2002. Moreover, these working time reductions were to take place without concomitant pay cuts. These economic policies have led Benjamin Barber to argue that 'Jospin's economic plan is the first in the Western world to recognize that employment, public sector work, and leisure are responsibilities of government no less pressing than those of the productivity and profitability of business' (Barber, 1998: 138). Confounding the negative forecasts of economic disaster resulting from working-time reductions, the Jospin government demonstrated that such a policy, coupled with high levels of productivity, could actually assist economic growth (although this is unlikely to harness widespread ecological support). Indeed, as we enter the new millennium France can proudly boast of the highest levels of economic growth in Western Europe, despite what neo-liberal economists would see as excessively *dirigiste* labour market policies, while simultaneously managing to reduce unemployment.

The move towards working time reductions or measures such as work-sharing has also been a long-term objective of the French Greens. Where countries such as the UK seem less likely to undertake radical work-sharing initiatives, France appears to be more willing to countenance these ideas:

> The question is not whether the work-sharing solution is economi-cally viable, since many reliable reports have now confirmed its

effectiveness, but whether it is acceptable to the key people concerned – workers and employers – and also politically accept-able, i.e. whether politicians are sufficiently convinced that there is political capital to be made from policies which require at least partial sacrifices. (Milner and Mouriaux, 1997: 62)

The Jospin government introduced its measures for working-time reductions to a 35-hour week against considerable opposition in France from, among others, employers, economists, President Chirac and some within the Socialist Party itself (Ardagh, 1999: 31–2). The common fears were raised that it would generate higher employment costs and indeed that should have been the case according to orthodox economic thinking. However, as Dumas (2000) notes, 'theoretical rigidity bends in practice'. Part of the reason for this was the fact that the 35 hour week policy, depicted as the only genuinely radical policy of a moderate government, was only enacted through offering various sweeteners to employers, including loopholes surrounding the possibil-ities for extended overtime if employers and employees agreed upon their desirability. Moreover, in the context of high levels of economic growth with rising tax revenues, the government was able to sweeten the pill by cuts in social security contributions and 'the relaxation of restrictive practices governing hours and functions at work employers have been seeking for decades' (Dumas, 2000).

This demonstrates how the policy of reduced working hours is not a zero-sum game but one that can open up a new range of innovative labour market policies to governments prepared to entertain such a measure. Of course the quid pro quo in the French example has been the appeasement of employers and sacrifices by the French labour movement. Whilst this may be palatable for some elements of the social democratic left, it is unlikely to find quite so much favour within Green circles especially in their more fundamentalist strands (Little, 2000). Nonetheless, despite the rhetoric of economic liberals, Jospin's example shows that radical options are open to contemporary govern-ments without economic downturn ensuing. Thus, many economic commentators have noted that the 'French economy, allegedly weighed down by over-regulation and high taxes, has emerged as the most dynamic of the large European Union economies. It now has fast growth, low inflation, trade in surplus and falling unemployment' (Tylecote, 2000: 21). That said, as Gorz (1999) has noted, working time reductions could also be used by unscrupulous employers to meet the perceived exigencies of flexible global markets. In itself, a policy of working time reductions is neither harmful nor beneficial. Moreover, the long-term impact of working time reductions on the French economy are not yet clear. Some economists are concerned that the extension of the policy to small businesses in 2002 will hinder flexibility in the future and undermine the current success story (Islam, 2000).

It should be clear that active labour market policies that address the need to regulate working hours are not, in themselves, economically harmful. In the context of contemporary France, working time reductions have been enacted within a political agenda that rejects any move towards Anglo-American models of flexibility and deregulation. In some circumstances they may prove to be highly beneficial to modern economies and this has silenced the kneejerk responses of Jospin's critics who saw the 35-hour week as a recipe for fewer jobs, less investment and declining competitiveness. At the same time as reduced working hours are not necessarily a panacea for economic problems such as unemployment, they are not inherently harmful either. In other words moves towards working time reductions will be workable or not depending upon a wider raft of social and economic policies. In the case of France the Jospin government sought to maintain wage levels but, as a safety net, also committed itself to the creation of 700,000 new jobs in the 'civil society' sector (Barber, 1998: 137). Barber notes that this strategy, which is targeted especially on the young unemployed (funded mainly by government with some input from the private sector), would provide status for the idea of public work. Thus, working time reductions are not only a possible strategy to manage economic affairs, but they also offer new opportunities for reimagining the very nature of work itself. The latter concern fits with the Green pursuit of the legitimation of unpaid work as a worthwhile and valuable form of activity. That said, Greens would be wary of a project that creates jobs (often paid at minimum wage levels) for the young as a targeted group, rather than redistributing work more equitably amongst the population as a whole.

Conclusion: Green dilemmas

It should be clear that there are good reasons why working time reductions might be attractive to Greens. The impetus behind a shift away from the primacy of paid work could make a significant contribution to the development of some kind of post-materialist culture. By reducing the centrality of paid work and, in so doing, attempting to promote non-instrumental attitudes towards work and time usage, potential emerges for Greens to encourage individuals to engage in more 'ownwork' (Robertson, 1989) and activities that do not have financial remuneration as their primary *raison d'être*. Despite this, four main problems have emerged from the Green perspective that make this potential harder to realise. One of these has been alluded to already: where a policy of working time reductions has emerged on the political agenda, it has tended to be at the behest of the left in the form of trade unions or socialist governments such as that of Jospin in France. In itself this is not problematic and there is no inherent reason why Greens should oppose strategies for redistributing paid work more equitably because they are part of a leftist agenda. However, socialist strategies for working time reductions such as that of Jospin have not been suggested or implemented as part of a fundamental challenge to global markets but rather as a response to the new flexibility required by them. Thus, whilst reduced working hours can be an alternative to the dominance of paid work in contemporary Western societies, leftist advocates are not necessarily attuned to the post-materialist agenda that might be endorsed by Green strategies. The continued domination of the demands of global markets appears likely to remain the context within which working time reductions are put forward by the left and this is unlikely to satiate the ecological appetite for radical changes in the global economy.

The second dilemma that Greens must address when examining working time reductions is whether social policies might take priority over economic policies such as reduced working hours. This feeds into the question of where social and economic policies of this kind should come on the Green agenda that is dominated by clearer ecological imperatives linked to the degradation of the ecosystem. This problem relates to the viability of proposals such as a guaranteed Basic Income and whether this should be prioritised over more complicated policies such as working time reductions which may be much more difficult to implement. One of the proposed advantages of guaranteed income schemes, as noted in Chapter 8, is their simplicity in terms of administration and delivery. Schemes that are related to reducing working hours are more likely to generate the kind of bureaucracy and state regulation that Greens traditionally oppose. This is not to say that systems of decentralised and local administration of working time reductions are impossible but that, in the context of modern economic relations, the need for a role for the state in enacting such a policy cannot be overlooked. In this scenario, despite the prime importance of working time reductions in the development of a sustainable long-term future, Greens may be more likely to support social policy measures in the shorter term. From the analysis above, however, it should be clear that a long-term strategy for a sustainable economy and affordable and efficient welfare regimes will be likely to involve a combination of a guaranteed income and working time reductions.

The third dilemma Greens must address is the degree of importance that should be attached to measures concerned with human welfare and economic arrangements. Of course all Greens are aware of the problems for the ecosystem that are generated by contemporary and past economic strategies. Nonetheless, the immediacy of environmental problems tend to dominate Green politics and there is a temptation within the Green movement to prioritise issues of pollution and adverse environmental impact over areas like working time reductions that appear as concerned with human well-being as they are with the welfare of the natural environment. It is worth reiterating that this does not mean that Greens should be unconcerned with reduced working hours but rather that there might be something of a lexical order in Green thinking whereby some issues are prioritised over others.

Thus, in terms of the economy, Greens might see fit to tackle companies or industries that pollute the environment before addressing issues such as the hours that the humans working therein perform. In this sense, ensuring that an appropriate economy emerges in a sustainable society may take precedence in Green thinking over the way humans work within industry and the economy. Ultimately any Green blueprint for a sustainable society will be enacted gradually and some policies will be implemented before others – in this context issues that are more directly harmful to the ecosystem are likely to be those that head the Green agenda. Of course, Green politicians must be expected to develop the full range of social and economic policies but the likelihood is that some of these will be prioritised and working time reductions are unlikely to be at the forefront of the agenda. The dilemma then for Greens is how to combine their concern for human welfare with environmental sustainability and the decision about priorities in that equation.

The final dilemma for Greens when addressing working time reductions is of a more philosophical nature. Ostensibly, there is much potential for Greens in the climate of contemporary politics where as much emphasis is placed upon responsibilities as rights (not least in the eyes of politicians such as Tony Blair). In this context opportunities exist for Greens to stress the obligations and responsibilities that we owe to the world around us and to future generations. However contemporary discourses are dominated by rights and responsibilities as reciprocal phenomena – we cannot enjoy the benefits of rights without taking on board the obligations that we reciprocally owe to one another and the state. This is problematic for Green politics because, as Dobson notes, Green theories of responsibility and obligation tend not to involve direct notions of reciprocity:

> the ecological contribution ... lies in its severing the mainstream connection between rights and obligations. The source of the ecological citizen's obligations does not lie in reciprocity or mutual advantage, but in a non-reciprocal sense of justice, or of compassion. The obligations that the ecological citizen has to future generations and other species ... cannot be based on reciprocity, by definition. Ecological citizens can expect nothing in return from future generations and other species for discharging their responsibilities towards them. Ecological citizenship's obligation is owed to strangers, who may be distant in time as well as space. (Dobson, 2000b: 218)

This helps to explain why working time reductions hold such an ambiguous position within the pantheon of Green political ideas. Whilst they can be regarded as a contribution to the development of a less materialist and less instrumentalist culture (that could form part of an obligation to future generations), in practice reduced working time is usually seen as a policy for the present. Thus the focus on obligations is usually presented as reciprocal: we should all work less to allow everyone to work. Clearly this moves us beyond a Green concern with 'a non-reciprocal sense of justice, or of compassion'. It suggests that we do have reciprocal obligations to one another in the production of social welfare and that Greens have not, nor should not, abandon reciprocity as part of their conceptualisation of justice and compassion. Whilst the future-oriented dimension of Green thinking is one of its greatest strengths, we must also recognise the currency of many Green proposals such as working time reductions. They reinforce the commitment to the common project of humanity to ensure human welfare within a broader context of environmental sustainability. Ultimately the latter is likely to have a lexical priority over social welfare measures in much Green thinking but this does not preclude the emergence of Green strategies for working time reductions through the process of compromise and negotiation with other sympathisers such as the libertarian left (see Chapters 2 and 3). In this sense there is no reason why the priority that Greens give to the natural environment should undermine the contribution that working time reductions could make to the creation of more sustainable forms of life and social welfare more generally.

11
Eco-Taxation in a Green Society
James Robertson

Introduction

This chapter outlines an approach to public finance and monetary policy which will help to shape the future of the welfare state. As well as being greener, it will also improve economic and social outcomes in line with the aims of an enabling state. Many of today's revenue sources and spending objectives encourage outcomes that are economically, socially and ecologically perverse. This new approach will involve changes in taxation, other sources of public revenue and in the objectives of public spending (Robertson, 1998, 1999, 2000; Brown, 2000). The new approach will recognise that – as Mason Gaffney, the distinguished professor of resource economics at the University of California, has pointed out – right-wing libertarian economists are wrong when they proclaim TANSTAAFL ('There Ain't No Such Thing As A Free Lunch') in reply to calls for economic justice, social justice and environmental sustainability. Nature and the activities of society as a whole provide many free lunches in the form of common resources. The real questions are 'who is *now* getting the free lunches' and 'who *should* get them'? In other words, how should the value of common resources be shared?

We start by examining the reasons why the present structure of taxation and public revenue needs to be changed. We then consider what the term 'common resources' includes and the extent to which the value of common resources can be captured as public revenue. In this context, we pay particular attention to the value of new money being put into circulation each year since the significance of this issue has usually been ignored. We then discuss whether environmental taxes and taxes on energy must necessarily have regressive effects, hitting poorer people harder than richer people. We conclude that they need not do so, provided that they are part of a package of changes (including Basic Income) capable of offsetting the potentially regressive effects of environmental taxes.

Turning then to public spending, we note that since large sums of taxpayers' money are now spent on perverse subsidies to business, pressure is likely to grow to reduce this expenditure and make it available for other uses. We also note that 'hypothecating' the revenue from environmental taxes and earmarking them to be spent as 'eco-bonuses' would help to offset any regressive effects which such taxes might otherwise have. Therefore, distributing that revenue as eco-bonuses to all citizens could be one of a number of transitional steps towards a BI.

We then outline how this approach to public revenue and public spending can be applied to the future development of global, as well as national, institutions of economic and financial governance. We briefly discuss how distribution and the role of market forces will be understood in a society in which a Green approach to social welfare plays a key part. We conclude with some remarks about the practicalities of the transition to Greener political economy and a Green welfare state.

The problems of public revenue

However rosy some governments', including the British government's, finances may appear to be at any particular time, it is widely recognised that the longer term pressures to restructure existing tax systems will continue to grow. These pressures include, for example:

- Today's taxes are becoming too complicated and expensive for tax authorities and taxpayers to manage satisfactorily.
- In an increasingly competitive global economy, the mobility of capital and highly qualified people will continue to press national governments to reduce taxes on incomes, profits and capital.
- In ageing societies, opposition will grow to taxing fewer people of working age on the fruits of their efforts in order to support a growing number of so-called 'economically inactive' people.
- As Internet trading (e-commerce) develops, it will become more difficult for governments to collect customs duties, value added tax and other taxes and levies on sales. This applies especially to sales of products and services that can be downloaded directly from the Internet – including music, films, pictures, games, and advice and

information of every kind. The Internet will also make it easier for businesses and people to shift their earnings and profits to low-tax regimes.

- International bodies like the OECD and the EU are demanding action against tax havens. In 1998 it was estimated that the £400 billion invested in Britain's tax havens – like the Channel Islands and Isle of Man – meant a tax loss of at least £20 billion a year to the UK exchequer. $6 trillion was estimated to be held in tax havens worldwide (*Guardian*, 1998). The results, apart from lost tax, include economic distortions and criminal money laundering on a massive scale. Although some changes are now being introduced, the best way to tackle the problem may be to shift taxation away from things that can migrate to tax havens, like incomes, profits and capital, and on to things which cannot migrate, such as land.

These growing pressures to shift the tax base reinforce the positive economic, social and environmental arguments for taxing 'bads', not 'goods' that have been widely discussed in recent years. Two American reports have spelled them out in an accessible way (Hamond, 1997; Durning and Bauman, 1998). The idea is to move the burden of taxes away from useful enterprise and employment on to the ownership and use of common resources, including land, energy and the capacity of the environment to absorb pollution. For example, the carbon/energy tax proposed by the EU in the 1990s would have used the revenues from taxes on fossil fuels to reduce the levels of tax on employment.

In short, the present tax structure in most countries is economically, socially, environmentally and ethically perverse for the following reasons:

- The taxes fall heavily on employment and on rewards for work and enterprise, but fall lightly on the use of common resources. So they encourage economic inefficiency in the use of resources of all kinds, i.e. over-use of natural resources and the under-use of human resources.
- In addition to their damaging economic, social and environmental effects, the taxes are unfair and illogical. They penalise 'value added' (the positive contributions people make to society) and they fail to penalise 'value subtracted'. Only exceptionally do they make people pay for using or monopolising common resources and thereby preventing other people from using them.

- They encourage and allow rich people and businesses to escape, or at least minimise, their tax obligations via such escape routes as tax havens and family trusts.

Common resources

These problems open up the need and the prospects for a new approach to fiscal policy, designed to collect the value of common resources as public revenue and to share it among all citizens.

Common resources are resources whose value is due to nature, the activities of society as a whole and not to the efforts or skill of individual people or organisations. Land is an obvious example. The value of a particular land-site, excluding the value of what has been built on it, is almost wholly determined by the activities and priorities of the society around it. For example, when the route of the Jubilee line in London was published, properties along the route jumped in value since access to them was going to be much improved. So, as a result of a public policy decision the owners of the properties received a windfall financial gain. They had done nothing for it and they had paid nothing for it but had been given a free lunch at taxpayers' expense. Calculations made in a New Economics Foundation report (Robertson, 1994,) based on 1990 values, suggested that the absence of a site-value tax on land might be costing up to £90 billion a year to UK taxpayers. This reflects an important failure, but only one of many, to collect the value of common resources as public revenue. By contrast, the auction of licences to use the radio spectrum for the third generation of mobile phones in Britain raised £22.5 billion for the government in 1999. It was followed by similar auctions in other countries such as Germany, Italy and Spain. That is a good example of the contribution which the value of common resources can make to public revenue.

Since it would take too long to run through each one, the following common resources can be identified as the most important:

- land (its site value);
- energy (its unextracted value);
- the environment's capacity to absorb pollution and waste;
- space for road traffic, air traffic, etc.;
- water for extraction, use and for waterborne traffic;
- the electro-magnetic (including radio) spectrum;
- genetic resources;
- the value arising from issuing new money (see below).

Since the aggregate annual value of the above is considerable, sharing it out among all citizens would go a long way to eliminating the need for many existing taxes.

To summarise, then, public revenue in the future is less likely to derive from the conventional taxes prevalent today and more likely to derive from taxes on the kind of common resources listed above. We will probably no longer tax people and businesses as heavily as now on what they *earn*; instead, we shall require them to pay for the value they *subtract* by their use or monopolisation of common resources. This change will be an essential feature of an environmentally sustainable economy. It will also make an important contribution to an economy that is more equitable, as well as more efficient. It will be part of the context for a greener welfare state.

To illustrate these points, let us look at the one common resource that is often unduly neglected.

Creating new money

At present in Britain less than 5 per cent of new money is issued and put into circulation by the government and the Bank of England as cash (coins and banknotes). The remaining 95 per cent of new money is non-cash money created and put into circulation by commercial banks. The banks simply create the money out of thin air by crediting it to the current accounts of their customers as interest-bearing, profit-making loans. A recent report, *Creating New Money*, has estimated that the annual loss of public revenue from allowing the banks to create non-cash money is about £45 billion (Huber and Robertson, 2000)[1]. To rectify this state of affairs Huber and myself have proposed a simple reform with two elements.

First, central banks should create the amount of new cash and non-cash money they decide is needed to increase the money supply and credit it to their governments as public revenue. Governments should then put it into circulation as public spending. In deciding how much new money to create, central banks should operate with a high degree of independence from their governments – as the Monetary Policy Committee of the Bank of England has done since 1997 – as it should not be made possible for politicians to 'debauch the currency', in Lenin's phrase, for electoral purposes. Secondly, it should be made illegal for anyone else to create new money denominated in the official currency. Commercial banks will then be excluded from money creation and will be limited to borrowing, but no longer creating, the money they need to lend.

This reform will restore 'seigniorage' in a form adapted to the conditions of the Information Age. That is to say, it will restore the prerogative of the state to issue money, and to capture as public revenue the income that arises from issuing it, in an age when most money has become electronic information rather than metal and paper tokens. Originally, seigniorage was the revenue enjoyed by monarchs and local rulers from minting coins. It reflected the fact that the coins were worth more than the costs of producing them. As over several centuries the physical characteristics of money have changed from metal to paper to electronics, and as banking practices have developed, the relative importance of that original source of seigniorage has gradually dwindled. Now that most money takes the form of electronic entries in computerised bank accounts, extending the traditional principle of seigniorage to non-cash money will correct the anomaly that has grown up over the years.

The arguments for this monetary reform are not limited to the contribution (of £45 billion in the UK) which it will make to public revenue, important though that will be. It will have additional beneficial social, environmental and economic effects. For example, it will tend to bring about lower interest rates and lower inflation, and it will tend to create greater economic stability by enabling the central bank to smooth out the peaks and troughs of business cycles more effectively than it can do today. Many of its advocates also argue that it will reduce the damaging social and environmental effects of rising levels of indebtedness caused by issuing new money into circulation in the form of interest-bearing debt. Moreover, some advocates of this monetary reform, going back to C.H. Douglas and the Social Credit movement of the 1920s and 1930s, propose that the revenue from it should be used to finance a National Dividend or, as we would call it today, a Basic Income. This proposal is supported by growing numbers of Green political thinkers and would make a direct contribution to social welfare policy, as Chapter 8 made clear.

The proposed monetary reform will also help to clarify monetary statistics, monetary definitions and monetary terminology, and make it easier for citizens and politicians of democratic countries to understand how the money system works and how it could be made to work better for the common good. This could play an important part in the greening of the welfare state. At present, it is difficult for politicians, officials, citizens and taxpayers to evaluate monetary and financial policies and policy options. In a democracy there should be wider understanding and discussion of how money created in accordance with the government's monetary policy should be spent into circulation. Should it be spent on achieving social and environmental objectives or should it be spent on loans for profit-making, private-sector business investment and private consumption? The outcomes of such a widening of public and political debate could be relevant to welfare policy. At all events, monetary reform would mark an important further step towards what, in his 1999 Mansion House speech, Chancellor Gordon Brown called 'transparency in policy-making, involving an open system of decision-making in both monetary and fiscal policy' (Brown, 1999).

Must eco-taxation be regressive?

Important potential new sources of public revenue would thus become available if it were decided to make people pay for the value of the common resources they use, rather than to tax them on the rewards they get for helping to meet society's needs and wants. The proposal to shift the burden of taxation to those new sources of revenue inevitably raises questions. One question is whether environmental taxes and resource taxes are bound to be regressive. A wide-ranging discussion of this and other aspects of 'a new politics of tax for public spending' was published in a recent Fabian Society report (Fabian Society, 2000).

If existing taxes on incomes, profits and savings were simply replaced with environmental taxes and resource taxes on consumers then they would hit poorer people relatively harder than richer ones. Almost regardless of the taxes they replace, energy taxes and other ecotaxes are bound to have this regressive effect if they are applied *downstream* at the point of consumption. For example:

- VAT on household energy hits poorer households harder than richer ones, because they do not have the money to pay the higher cost of the tax or to invest in greater energy efficiency in order to reduce the tax;
- similarly, fees and charges to reduce urban congestion will hurt small trades-people who need to use their vehicles for their work, but will be painlessly absorbed by users of chauffeur-driven limousines.

How is this problem to be solved? Hitherto, the question has been considered case by case as it applies specifically in detail to each of the growing number of new taxes introduced or proposed for purposes of resource conservation and environmental protection. At the time of writing, in the UK, those include: landfill tax, VAT on domestic energy consumption, climate change levy, aggregates tax, fuel duties, vehicle excise duty, company car taxes, road congestion and workplace car-parking taxes (Fabian Society, 2000: 299). However, if resource taxes and ecotaxes are to replace existing taxes significantly, it is also necessary to consider some points of more general application. These have a bearing on the future development of the welfare state.

First, ecotaxes should, as far as possible, be applied *upstream*. Of key importance will be a tax on carbon-energy (or on fossil fuels and nuclear energy) *collected at source*, cascading down through the economy and raising the cost of the energy content of all goods and services. It would reduce pollution because pollution arises predominantly from energy-intensive activities. It would be administratively simpler and easier to understand than a proliferation of separate eco-taxes imposed directly on individual consumers and polluters. And, by clearly raising costs for companies which extract fossil-fuel energy, or generate nuclear energy, or which produce energy-intensive goods and services, it will be seen to hit richer people's incomes by reducing the salaries, dividends, capital appreciation, etc. derived from energy-intensive production. It would thus be seen to be fairer and less discriminatory than taxes that fall directly on consumers at the point of sale. Even so, however, and depending on the elasticity of the prices that consumers are able or prepared or forced to pay for using energy and continuing to pollute, the regressive effect of energy taxes and other ecotaxes will still tend to operate unless it is compensated for in other ways. In other words, resource taxes and other environmental taxes need to be considered, not as separate items each on its own, but as parts of a larger package whose overall impact is progressive and not regressive.

Secondly, then, one conclusion to be drawn is that if resource taxes and ecotaxes are to be introduced they should include a *site-value tax on land*. That is, a resource tax that is clearly progressive. No one supposes it is the poor who are enriched by the private 'enclosure' of the value of land. There are also other good reasons for treating a site-value tax as an environmental tax and as a welfare measure. It can discourage suburban sprawl, urban decay and the disproportionately rising house prices that poorer people cannot afford to pay (Durning and Bauman, 1998: 57–65).

A third way to counter the regressive effects noted above is by *hypothecation*: to earmark the revenue eco-taxes so that they are spent on specifically progressive purposes. For instance, a 1994 German study concluded that if part of the revenue from an energy tax were distributed to households as an eco-bonus, the change would not only have positive economic and employment effects, but would also reduce the net tax burden on low-income households (DIW, 1994). In another example, a 1989 Swiss study concluded that if the revenue from levying two Swiss francs per litre of petrol were distributed to all adults as an eco-bonus, people driving less than 7,000 km a year – including people who did not have a car – would benefit, while people driving more than 7,000 km would lose (Weizsacker, 1994: 76).

Since there are different kinds of hypothecation, and there are arguments for and against them all (see Fabian Society, 2000: 154–85), we do not have to go into them in detail here. In the context of greening the welfare state, though, the relevant questions are: could eco-bonuses contribute to a Basic Income and could a Basic Income be financed from resource tax revenues? Quite simply, if a democratically elected government so decided it could earmark the revenue from resource taxes and pollution taxes as contributions towards a partial Basic Income (see Chapter 8) and this could, in fact, be a transitional step towards a full Basic Income.

Public spending

The following three aspects of public spending are among those likely to rise up the agenda of political and public debate in the coming years. They are all relevant to the greening of welfare.

The first is about the payment of taxpayers' money to business corporations. Governments now spend very large sums in contracts, subsidies, inducements, incentives and various other contributions to corporate budgets. To take subsidies alone, it is estimated that, worldwide, $1.5 trillion is spent every year on perverse subsidies, perverse in the sense of having economically, socially and environmentally damaging effects (Myers, 1998). Public pressure seems certain to grow on members of parliaments and other opinion formers to reduce the amounts of taxpayers' money that governments spend in this way – thus making more available for public spending on other things.

The second point has been mentioned already: hypothecation. In recent years, hypothecation has been discussed with particular relevance to environmental taxes. For example, congestion charges on motor transport in cities are expected to be more readily accepted if the revenue from them is clearly spent on ways that improve public transport. In November 2000, however, it was seriously proposed, though with acknowledged reservations about the difficulties, that half of the revenue raised from income tax in the UK should be earmarked for spending on the National Health Service (Fabian Society, 2000: 154–85). Whatever the short-term public and political reaction turns out to be, this particular proposal seems unlikely to make a positive contribution to a greener welfare state. It would make clearer sense to earmark other taxes than income tax for spending on the NHS, such as taxes on tobacco, alcohol, road travel and transport, and other products and activities involving health risks and costs.

The third point is the probable further extension of benefit payments or tax credits to various categories of people. The Working Families Tax Credit, introduced in 1999 in the UK, has led on to a debate about how to guarantee the incomes of other active citizens such as carers (for example by paying their national insurance), parents and social entrepreneurs. Combining developments like these with eco-bonuses could possibly lead towards a universal Basic Income, support for which seems to be growing. As Chapter 8 noted, a Basic Income would reflect the idea that some of the public revenue arising from the value of common resources should be shared equally and directly among all citizens, leaving only some of it to be spent on their behalf by government officials and businesses via public spending programmes.

The global dimension

The principles underlying what this chapter has said about national public finance and the creation of new money apply at the global level too. The Commission on Global Governance recognised in 1995 that global taxation is needed 'to service the needs of the global neighbourhood' (Commission on Global Governance, 1995: 217–21). Global taxes, based on the use each nation makes of global commons, could include:

- taxes and charges on use of international resources such as ocean fishing, sea-bed mining, sea lanes, flight lanes, outer space and the electro-magnetic spectrum;
- taxes and charges on activities that pollute and damage the global environment, or that cause hazards across or outside national boundaries, such as emissions of CO_2 and CFCs, oil spills, dumping wastes at sea and other forms of marine and air pollution.

The Commission also pointed out that international monetary reform is becoming urgent: 'A growing world economy requires constant enlargement of international liquidity' (Commission on Global Governance, 1995: 186).

In *Creating New Money* Huber and I suggested that the principle underlying seigniorage reform at the national level could be applied at the global level too (Huber and Robertson, 2000: 56–58). Revenue from global taxes and from global seigniorage could then provide a stable source of finance for UN expenditures, including its peace-keeping programmes. Moreover, some of the revenue might be distributed to all nations according to their populations, reflecting the right of every person in the world to a global Basic Income based on a more equal share of the value of global resources.

Such reforms as this would:

- encourage sustainable development worldwide;
- generate a much needed source of revenue for the UN;
- provide substantial financial transfers to developing countries by right and without strings, as payments by the rich countries for their disproportionate use of world resources;
- help to liberate developing countries from their present dependence on aid, foreign loans and institutions like the World Bank and the International Monetary Fund which are dominated by the rich countries;
- reduce the risk of another Third World debt crisis;
- recognise the shared status of all human beings as citizens of the world.

In other words, the development of new fiscal and monetary arrangements based on a more equal sharing of common resources can make an important contribution to welfare policy at the global level, as well as at the national level (International Union for Land Value Taxation, 2000). Yet what this new welfare policy and Green political economy would require is a new understanding of the meaning of distribution and the role of market forces.

Today the welfare state is involved in large-scale redistribution to counteract the social damage caused by the outcomes of economic activity. By contrast, Green political economy calls for *pre*distribution, based on sharing the value of common resources.[2] Whereas redistribution aims to correct the outcomes of economic activity after the event, predistribution will involve sharing the value of essential inputs to economic activity; it will act at the beginning of the economic 'pipe' instead of the end of it. Whereas redistribution reinforces dependency, predistribution will be enabling. Because it will address underlying causes of economic injustice, inequality and exclusion, predistribution will be an essential feature of a prosperous economy in an inclusive society. It will reverse the private 'enclosure' of common resources and the private enjoyment of 'free lunches'.

A Green political economy will also recognise that a completely free market economy is conceptually impossible. Even if it were possible to start with a totally blank economic page, some people and businesses would soon begin using their freedom to diminish the freedom of others less powerful than themselves. The free economy would soon become unfree.

At a more sophisticated level of understanding, the governments of countries like the UK account for roughly 35–40 per cent of GDP. They take it out of the market economy as taxation and then put it back into the market economy as public spending. This has a massive impact on relative costs and prices throughout the economy, with taxes adding to the cost of everything that is taxed and public spending reducing the cost of everything it supports. In other words, it is obvious that the proverbial 'level playing field' is a mirage, a conceptual impossibility. The total composition of public revenue combined with the total composition of public spending will always provide a framework for the economy which skews its price structure some ways rather than others.

To recognise this is not to argue in favour of a command economy based on detailed central planning by governments, or in favour of ongoing *ad hoc* interventions by governments in the workings of the market economy. It is to argue that the central aim of governments' financial and monetary policies must be to provide a framework of public revenue collection and public spending that predisposes the economy, operating as freely as possible within that politically designed and determined framework, to produce outcomes which broadly accord with democratically decided choices and preferences. This may be seen as a Green and socially responsive market economy, which differs in important respects from any of the conventional economies of the 20th century, capitalist, socialist or mixed.

Concluding remarks

Growing numbers of people around the world share a vision of a future less dominated by big business, big finance and big government than the globalised society of the late 20th century. Features of a new more people-centred and ecologically orientated society could include the following:

- Its citizens would be more equal in esteem, capability and material conditions of life than we are now.
- They would find it easier to get paid work. But they would no longer be so dependent on employers to organise it and provide their incomes as we are now.
- The industrial-age class division between employers and employees would continue to fade – as the old master/slave and lord/serf relationships of ancient and medieval societies have faded. It would become normal for people to work for themselves and for one another. Public policies would enable them to manage their own working lives.
- In exchange for their right to share in the value of common resources, people would expect to take greater responsibility for themselves and for the well-being of their families, neighbourhoods and society.

This vision of the future, and its implications for a greener welfare state, will seem utopian to many people today. But there is a tendency, as independent voices spread awareness of the need and possibilities for change, for mainstream opinion to shift after a certain time lag. The growing prominence of the environment on mainstream policy agendas over the past twenty years is a good example of this. As forward thinkers move ahead and mainstream opinion moves to catch up, no hard-and-fast boundary line separates the policy implications of what some see as utopian thinking from the actual political agenda as it evolves (Robertson, 1999: 2–3). This chapter has suggested that a reconstruction of public finance and the monetary system could be part of the greening of the welfare state. How, then, will the political will be mobilised to take things forward?

It is always difficult for established people pursuing a professional career in a particular walk of life to support proposals for its radical reform until there is widespread acceptance that they should. It may be unrealistic, therefore, to hope that in the immediate future financial and monetary reconstruction will attract support from the majority of today's experts and professionals in the welfare field. But, although they should not be expected to welcome it as desirable, they would nonetheless be wise to recognise it as a possibility that may help to shape the future of the welfare state.

The likelihood is that pressure for the changes discussed here will continue to build up from a wide range of sources in many countries. These will include government departments and agencies, academic faculties and research institutes, and non-governmental organisations (NGOs) and pressure groups. They are likely to be concerned with a wide range of issues, including taxation and public spending, banking and the monetary system, the environment and sustainable development, welfare and poverty, the Third World, globalisation, the international economy and international finance. As attention is increasingly focused on how the value of common resources is shared, pressure for 'joined-up' policy-analysis, research and campaigning may encourage more government departments, academic disciplines and NGOs to explore the topic jointly. Then, in due course, growing media interest and public concern may be expected to persuade mainstream politicians that the question has an important bearing on the political values they promote and the practical policies they adopt.

Notes

1   The New Environment of Welfare

1. 'Environmentalism' and 'ecologism' will be treated as synonymous for the reasons explained below.
2. Isaac Asimov once wrote a short story in which the last remaining animals are killed in order to create space for one more human.
3. This does not mean that animals should be entitled to claim benefits or sit qualifications! It does mean, though, that consideration for non-human life should be a stronger component of the decision-making process than at present.

2   The Ethical Foundations of a Sustainable Society

1. For the sake of consistency I have used the term 'interspecies justice' to describe the ethical principles operative in relation to human–nonhuman interaction. However, the term 'justice' should be understood rhetorically and as a shorthand label, since I have some misgivings about either the necessity and/or desirability of extending the discourse of social justice to cover human–nonhuman relations, the details of which I will not go into here. However, one argument is that while it is necessarily the case that non-humans and the natural world should be considered morally considerable, this does not mean that the only or most appropriate moral language or discourse we should use is that of justice (see Barry, 1999a). For a defence of extending justice to the nonhuman world, see Baxter (1999) and Dobson (1999).
2. In many respects the distinctions drawn between use/abuse use/non-use are similar to Kantian concerns relating to treating humans always as 'ends in themselves' and never purely as means. The Kantian injunction is not that we ought to never treat humans (or at least rational ones) as means, but rather than we should always regard them as ends in themselves as well. This implies that any 'use' of one human by another (for example, employment) ought to be tempered by moral considerations not related to that use. In a comparable fashion, an ethics of use seeks to persuade us that some human uses of nature ought to be governed by wider human interests which temper narrow considerations of economic efficiency, for example.
3. It is interesting to note that this formula of unequal distribution of economic wealth as necessary (and normatively justified) to produce maximum overall economic growth/wealth, is the dominant way in which social or distributive justice has been thought about for over 200 years, from Adam Smith's dictum to John Rawls' (1972) theory of justice. It is only in contemporary Green political economy and theory that we find attempts to work out what social justice may look like and require if divorced from economic

growth. One of the classic and most eloquent attempts to sketch what a 'post-growth' society and associated view of social justice may look like is John Stuart Mill's outline of the 'stationary state' economy (Mill, 1900).
4. One interpretation of the critique of orthodox economic growth is that economic security rather than economic affluence is important for a more equal social order within modern democracies. The Green view is that it is the distribution of wealth within society, not the absolute level of wealth, which is important in a democratic political system. Similarly, the lessons for social policy are that what is important is the distribution of work (not just waged employment in the formal economy), free time and other 'public good' dimensions of a decent quality of life, such as quiet, pleasant, clean work and living environments and personal security. Whereas traditional social policy has been concerned with the distribution of consumption, a green or more sustainable-orientated social policy is centrally concerned with the distribution of productive opportunities (including forms of self-provisioning). See Barry (1998) and Fitzpatrick (1998).
5. A further argument is given by Fitzpatrick (1998: 10) in favour of the positive benefits of a less growth orientated society and welfare state: 'A welfare state which depends upon growth ... is not efficacious because what is created is a ambulatory system of welfare which tends to treat the symptoms rather than the causes of social problems, i.e. it can lighten the stressful demands of too much work, consumption and competition, but it does little to prevent such alienating effects from being created in the first place.' Just as with decreasing inequality, there is a strong argument that to decrease many of the stresses of modern life and improve people's well-being requires breaking with economic growth as an integral part of achieving this.
6. It is of course a moot point whether the small variations to a basic and dominant materialist and consumption-based view of the good life, actually constitutes 'variety'. Does a view of the good which has as central aspects the accumulation of x amount of wealth, property, and the consumption and ownership of commodities differ in kind as opposed to degree to one which has $2x$ or $x-1$ amounts or degrees of these things as central? My own view is that this is just a variation on a theme rather than somehow constituting a different view of the good.

3 The Ideologies of Green Welfare

1. Vincent (1993a: 248) by contrast stresses the 'internal complexity' of ecologism.
2. See Freeden (1998) for an account of what is required for qualification as a 'full' ideology. This is discussed further below.
3. See for example Vincent (1993a, b), Hayward (1994), Freeden (1996), Stavrakakis (1997, 2000), Wissenburg (1997), Barry (1999a), Baxter (1999), Blühdorn (2000), Dobson (2000a, b). As for primary rather than secondary texts the academic literature on the philosophy of ecologism and the disputes between deep ecologists, social ecologists, ecofeminists, biocentrists and so on is huge. There is another literature that might be considered 'ideological' in that it seeks to appeal to a mass lay audience and gather support

for the cause. In addition to the direct output of Green political parties and pressure groups this would include Spretnak and Capra (1985), Porritt (1984, 2000), Porritt and Winner (1988), Irvine and Ponton (1989), Kemp and Wall (1990), McKibben (1990), Wall (1990), Day (1991), Pearce (1991), Tokar (1992), Jacobs (1996, 1999).

4. 'Ideology' has to be one of the most used and abused concepts in social science, and the theoretical literature regarding what ideology 'is' is sufficiently vast that justice could not possibly be done to it here. One could consult Larrain (1979), Thompson (1984), Hamilton (1987), Boudon (1989), Eagleton (1991), McLellan (1995), Freeden (1996), van Dijk (1998). These texts operate at various levels of sophistication and for students new to the subject I would suggest McLellan is the best place to start. What follows in this section is a purely schematic overview.

5. It is worth noting that although this text is rather dated now, it still has currency. When Stavrakakis contacted the UK Green Party to discuss their ideological perspectives for his 1997 paper, he was told to go and read *Seeing Green* (Stavrakakis, 1997: 277, fn51).

6. The history of this tradition is traced in Terry Eagleton (1991).

7. It should be noted that there are questions about the consistency with which Marx and Engels use the term 'ideology', although this need not concern us here (see Eagleton, 1991: Ch.3).

8. Indeed it seems to me a particular advantage of Freeden's approach that he leaves such questions open to empirical enquiry rather than as assumed elements of ideological thought. He does not, either, assume Dobson's functional trilogy of critique, transition, and utopia.

9. Freeden is at pains to point out that this is not an essentialist theory of an ideological core, but rather an empirical one.

10. This empirical approach does raise the problem of the necessary and sufficient conditions required for a text to count as a member of a particular ideological family. Is it necessary for a writer or party to self-identify with a particular ideology? If so, how would we categorise a group like the Russian 'Liberal-Democratic Party' who appear to be neither liberal nor democratic in outlook? Freeden's approach seems to consist in the following. An ideological tradition is developed initially through the work of people who self-identify with a particular ideological view. However, once that tradition begins to accumulate textual reference points, self-identification becomes less important, as by now observers can begin to discern the outline of the conceptual commitments that make up the ideological system. Thus it becomes possible to judge externally whether a particular text belongs within a particular ideological tradition. There are doubtless methodological problems with this approach, particularly with respect to retrospective judgements, but it nonetheless has an intuitive plausibility.

11. Compare his account of liberalism, which has 12 elements to the conceptual core (1996: chs 4–7). It is also worth comparing Freeden's account of the Green conceptual core with Blüdhorn's (2000: 18) similar exercise. The latter has environmental degradation, finiteness of resources, limits to growth, future ecological collapse, and global affectedness. The lack of fit between these two accounts might suggest a level of complexity in Green political thought unacknowledged by Freeden. It might also, however, be

due to these two characterisations operating at different levels of abstraction.

12. I follow Dobson in making this distinction (though see Chapter 1). For him, *environmentalism* entails a managerial approach to environmental problems, 'secure in the belief that they can be solved without fundamental changes in present values or patterns of production and consumption'. By contrast, *ecologism* 'presupposes radical changes in our relationship with the nonhuman natural world, and in our mode of social and political life' (Dobson, 2000a: 2).

13. The main source for this part of the chapter is the UK Green Party's "Manifesto for a Sustainable Society" (http://www.Greenparty.org.uk/homepage/policy/mfss/economy.html) and declaration of philosophical principles (http://www.Greenparty.org.uk/homepage/policy/mfss/pbasis.html). However, there are of course enormous problems of mapping between programmatic declarations of particular parties at particular moments on to historically evolved ideological families. Thus a number of self-declared Green texts will also be included, as will evidence from the German Green Party manifesto (http://www.gruene.de/sache/english/election98).

14. The UK Green Party subscribe to the ecocentric principle that 'value ultimately lies in the well-being of the whole ecosystem' and that humanity is dependent upon the natural environment. There is strong survey-based evidence that most UK Green Party members hold this ecocentric view, see Bennie *et al.* (1995).

15. In accord with Dobson's distinction between environmentalism and ecologism, see note 12.

16. The German Green Party are committed to 'needs-oriented basic support' sufficient to 'enable citizens to maintain their dignity and take a full part in social and cultural life' (for reference, see note 13). Kemp and Wall (1990: 102) note that a Basic Income will allow people the advantage to 'choose the age at which they retire'. Porritt (1984: 137) is also committed to the idea.

17. Although it might be fair to suggest that it is not uniquely determined, in that there may be other possible ways to go down this route to sustainability. See Chapter 8.

18. On bioregionalism see Sale (1985) and McGinnis (1999).

19. David Miller has remarked to me that Green ideology has single-handedly revived the heretofore defunct politics of anarchism – see Miller (1984) for his own discussion of anarchism.

20. Although Naess does wonder whether his view is 'too pessimistic'.

21. For an extended critique of Green ideology from this perspective see McHallam (1991).

22. Although given Freeden's (1996: 162) theoretical position with regard to the inevitability of hybridised variants of established ideological patterns, it might be suggested that he is not entirely consistent on this point.

23. See for example Robyn Eckersley's (1992: ch. 1) positioning of Green political thought. Bramwell (1989) challenges the view that ecologism has been, historically, a left-wing political position.

24. See John Gray (1993: Ch. 4); on attempts by the far right to co-opt ecologism see Coates (1993).

25. In particular, I would resist Dobson's contention that ecocentrism is a necessary condition of a text qualifying as ecological. This understanding excludes works such as Hayward (1994) and Barry (1999a) from consideration, even though they clearly arguing for far more than technocentric environmentalism. I would instead suggest that a *critical attitude to anthropocentrism* is a necessary condition of ecologism. This does not rule out the possibility that a particular version of anthropocentrism will be endorsed. I anyway believe that the ecocentric/anthropocentric divide is attributed far more import than it deserves in Green literature, a point I will be explaining at length in a forthcoming book (see Chapter 2 also).
26. Limits in the sense that, although ideologies might be able to contain some inconsistent or even contradictory elements and still gather popular support, it is assumed there are limits to these possibilities when the contradictions become manifestly obvious.
27. Certainly free-market environmentalists see the incentives of capitalism operating differently – although their arguments run into difficulties on the reasonable assumption of capital mobility
28. See Goodin (1992). He posits a radical distinction between the 'Green theory of value' on the one hand, and the 'Green theory of agency' on the other. It should be noted that this hard distinction between ends and means has been the object of considerable criticism.
29. Remember the UK Green Party contention that ensuring the material security of everybody was the *only* way of securing ecological sustainability.

4   Green Democracy and Ecosocial Welfare

1. Though there is disagreement as to whether such limits are fixed and ahistorical or malleable and historically relative (see Benton, 1996).
2. This argument resembles one made by John Barry in Chapter 2.
3. Although some have made a valiant effort to define a Green conservatism, e.g. Gray (1993), there is little reason to believe that this would not display the usual conservative trait of being subordinate to either economic liberal precepts or else social democratic ones. Witness Gray's (1998) argument that the best we can hope for is to defend local traditions against the inexorability of free market globalisation.
4. So, Greens need not 'disqualify' consumer preferences, as Wissenburg (1998: 221) suggests they do, but merely regard them as less conducive to the ecological common good.
5. And these schemes should obviously be made far more accessible than they are at present to those on low incomes through some kind of redistributive credit scheme.

5   Local Welfare: State and Society

1. I am grateful to Tony Fitzpatrick for extremely helpful comments on a draft of this chapter.

6 Quality of Life, Sustainability and Economic Growth

1. Daly (1996: 3) for example cites Mill as arguing for a 'stationary condition of capital and population' in which there would be more likelihood of 'improving the art of living ... when minds ceased to be engrossed by the art of getting on'.
2. I do not have a reference for this quote and would welcome the help of anyone who can supply one.
3. In fact, the original MEW was less concerned with the environmental costs of economic growth than the importance of social and economic trends which were unaccounted for within GDP.
4. In some of these studies – for example the Australian study and later versions of the US study – the adjusted index has been 're-branded' as the Genuine Progress Indicator, and includes some additional contributions not incorporated into the ISEW. However, the basic methodology is broadly comparable.
5. See note 2.
6. Each of Max Neef's nine needs exists in four 'existential modes': being, having, doing, and interacting.
7. The designation of subsistence and protection needs as 'material' and other needs as 'non-material' is in some sense an abuse of language, since it is the satisfiers rather than the needs themselves that embody materiality. Furthermore, almost any kind of activity implies the need for some kind of material inputs at some level. Nonetheless, the distinction is valid in the sense that nonmaterial needs could, in principle, be met with vanishingly small material inputs, even though in practice, modern society attempts to meet these needs in increasingly material ways.
8. See note 2.

8 With No Strings Attached? Basic Income and the Greening of Security

1. The proposal still goes under a number of different guises, e.g. Citizen's Income in the UK, but 'Basic Income' is probably the most widely used and recognised.
2. Many thanks to Bart Coenan of the Flemish Green party, Outi Hannula of the Green League of Finland and Hans Bakker of the Dutch Green party and Peter Romilly of the Scottish Green party.

9 The Social Economy and LETS

1. Between November 1998 and April 1999, in-depth action research (Stringer, 1996) was conducted with this LETS. The first stage comprised an initial survey of the 326 members to identify the character of the membership, trading levels and members' perceptions of the effectiveness of LETS in promoting social inclusion. The second stage used a multi-method approach including in-depth interviewing, focus groups and the researchers' participation in all aspects of the scheme to explore the multiple meanings of participation, visions of LETS development, and the barriers to increased

participation. In total, 29 in-depth interviews were completed and transcribed, with transcripts being returned to the interviewee for final edit and agreement for use in this research. Five focus groups were moderated with numbers varying between 4 and 7 participants (excluding the facilitator), again these were fully transcribed and returned to participants for final edit, and this process was evaluated by the participants.

11   Eco-taxation in a Green Society

1. *Creating New Money* is about how new money is issued and is not about new currencies like LETS. Those are important innovations, but different (see Chapter 9).
2. I owe this distinction to Joseph Huber, co-author of *Creating New Money*. See Fitzpatrick (1996) also.

Bibliography

Achterberg, W. (1999) 'From Sustainability to Basic Income', in Kenny, M. and Meadowcroft, J. (eds), *Planning Sustainability*, London: Macmillan.

Ackerman, B. and Alstott, A. (1999) *The Stakeholder Society*, New Haven, Conn.: Yale University Press.

Alcock, P. (1997) *Understanding Poverty*, London: Macmillan.

Allison, L. (1991) *Ecology and Utility: the Philosophical Dilemmas of Planetary Management*, Leicester: Leicester University Press.

Althusser, L. (1969) *For Marx*, London: Allen Lane.

Anderson, E. (2000) 'Optional Freedoms', *Boston Review*, 25(4).

Anderson, V. (1991) *Alternative Economic Indicators*, London: Routledge.

Andersson, J. O. (1996) 'Fundamental Values for a Third Left', *New Left Review*, 216.

Archibugi, F. and Koenig-Archibugi, M. (1995) *Industrial Relations and the Social Economy: Forms and Methods of 'Negotiated Destatalization' of the Social Welfare Systems in the European Union*, CEC: DG V.

Ardagh, J. (1999) *France in the New Century: Portrait of a Changing Society*, London: Viking.

Arendt, H. (1958) *The Human Condition*, Chicago: University of Chicago Press.

Atkinson, A. (1983) *The Economics of Inequality*, 2nd edn, Oxford: Oxford University Press.

Bahro, R. (1986) *Building the Green Movement*, London: GMP Publishers.

Barber, B. (1998) *A Place for Us: How to Make Society Civil and Democracy Strong*, New York: Hill & Wang.

Barry, B. (1995) *Justice as Impartiality*, Oxford: Oxford University Press.

Barry, J. (1994) 'Beyond the Shallow and the Deep: Green Politics, Philosophy and Praxis', *Environmental Politics*, 3(3).

Barry, J. (1998) 'Social Policy and Social Movements: Ecology and Social Policy', in N. Ellison and C. Pierson (eds), *Developments in British Social Policy*, London: Macmillan.

Barry, J. (1999a) *Rethinking Green Politics: Nature, Virtue and Progress*, London: Sage.

Barry, J. (1999b) *Environment and Social Theory*, London: Routledge.

Barry, J. and Proops, J. (2000) *Citizenship, Sustainability and Environmental Research*, Aldershot: Edward Elgar.

Baxter, B. (1999) *Ecologism: An Introduction*, Edinburgh: Edinburgh University Press.

Beck, U. (1992) *Risk Society*, London: Sage.

Beck, U. (1995) *Ecological Politics in an Age of Risk*, Cambridge: Polity Press.

Beck, U. (2000) *The Brave New World of Work*, Cambridge: Polity Press.

Bell, D. (1962) *The End of Ideology*, New York: Free Press.

Bell, D. (1973) *The Coming of Post-Industrial Society*, New York: Basic Books.

Bennett, F. and Walker, R. (1998) *Working with Work: an Initial Assessment of Welfare to Work*, York: Joseph Rowntree Foundation.

Bennie, G., Franklin, M. and Rudig, W. (1995) 'The Ideology of the British Greens', in Rudig, W. (ed.), *Green Politics Three*, Edinburgh: Edinburgh University Press.

Benton, T. (1993) *Natural Relations*, London: Verso.

Benton, T. (ed.) (1996) *The Greening of Marxism*, New York: Guilford Press.

Bhatti, M. (1996) 'Housing and environmental quality in the UK', *Policy and Politics*, 24.

Blowers, A. and Glasbergen, P. (1995) 'The search for sustainable development', in P. Glasbergen and A. Blowers (eds), *Perspectives on Environmental Problems*, London: Edward Arnold.

Bludhorn, I. (2000) *Post-Ecologist Politics*, London: Routledge.

Bookchin, M. (1972) *Post-Scarcity Anarchism*, Montreal: Black Rose Books.

Bookchin, M. (1994) *Which Way for the Ecology Movement*, Edinburgh: AK Press.

Boudon, R. (1989) *The Analysis of Ideology*, Cambridge: Polity Press.

Bradshaw, J. & Hutton, S. (1983) 'Social policy options and fuel poverty', *Journal of Economic Psychology*, 3.

Bramwell, A. (1989) *Ecology in the 20th Century: a History*, New Haven, Conn.: Yale University Press.

British Green Party (1995) *A Guide to the Green Party's Basic Income Scheme*, London: Green Party.

British Green Party (1997) *General Election Campaign Manifesto*, London: Green Party.

Brown, G. (1999) *Mansion House Speech* (www.hm_treasury.gov.uk/press/1999/p94-99.html).

Brown, G. (2000) *James Meade Memorial Lecture* (www.hm_treasury.gov.uk/speech/cx80500.html).

Brundtland Commission (1987) *Our Common Future*, Oxford: Oxford University Press.

Bulmer, M., Lewis, J. & Piachaud, D. (eds) (1989) *The Goals of Social Policy*, London: Unwin Hyman.

Cahill, M. (2001) *The Environment and Social Policy*, London: Routledge.

Cahill, M. (1991) 'The Greening of Social Policy?', in N. Manning (ed.), *Social Policy Review 1990–91*, Essex: Longman.

Cahill, M. (1999) 'The Environment and Green Social Policy', in J. Baldock, N. Manning, S. Miller and S. Vickerstaff (eds) *Social Policy*, Oxford: Oxford University Press.

Cahill, M. and Fitzpatrick, T. (eds) (2001) *Environmental Issues and Social Welfare*, special edition of *Social Policy & Administration*.

Carter, A. (1999) *A Radical Green Political Theory*, London: Routledge.

Castaneda, B. (1999) 'An Index of Sustainable Economic Welfare for Chile', *Ecological Economics*, 28(2).

Chanan, G. (1999) 'Employment and the social economy: promise and misconceptions', *Local Economy*, 13(4).

Christie, I. and Warpole, K. (2001) 'Quality of Life', in A. Harvey (ed.), *Transforming Britain*, London: Fabian Society.

Christoff, P. (1996) 'Ecological Modernisation, Ecological Modernities', *Environmental Politics*, 5(3).

Clasen, J. (ed.) (1998) *Social Insurance in Europe*, Bristol: Policy Press.

Coates, I. (1993) 'A Cuckoo in the Nest: the National Front and Green Ideology', in Holder, J. (ed.), *Perspectives on the Environment*, Aldershot: Avebury.

Cobb, C. and Cobb, J. (eds) (1994) *The Green National Product*, Lanham: University of America Press.

Cohen, J. and Rogers, J. (1995) *Associations and Democracy*, London: Verso.

Commission on Global Governance (1995) *Our Global Neighbourhood*, Oxford: Oxford University Press.

Community Development Foundation (1995) *Added Value and Changing Values: Community Involvement in Urban Regeneration: a 12 Country Study for the European Union*, final report CEC: DG XVI.

Compston, H. (ed.) (1997) *The New Politics of Unemployment: Radical Policy Initiatives in Western Europe*, London: Routledge.

Connelly, J. and Smith, G. (1999) *Politics and the Environment: from Theory to Practice*,. London: Routledge.

Cook, C. (1976) 'Labour and the Downfall of the Liberal Party, 1906-14', in A. Sked and C. Cook (eds), *Crisis and Controversy*, London: Macmillan.

Cook, D. (1997) *Poverty, Crime and Punishment*, London: Child Poverty Action Group.

Dalton, H. (1920) 'The measurement of the inequality of incomes', *Economic Journal*, 30.

Daly, H. (1996) *Beyond Growth: the Economics of Sustainable Development*, Boston: Beacon Press.

Daly, H. and Cobb, J. (1990) *For the Common Good*, London: Green Print.

Darier, E. (ed.) (1999) *Discourses of the Environment*, Oxford: Blackwell.

Day, D. (1991) *The Eco Wars*, London: Paladin.

Dean, H. & Melrose, M. (1996) 'Unravelling citizenship: the significance of social security benefit fraud', *Critical Social Policy*, 16.

Dean, M. (1999) *Governmentality*, London: Sage.

De-Shalit, A. (2000) *The Environment: Between Theory and Practice*, Oxford: Oxford University Press.

DETR (1998) *Community-based Regeneration Initiatives: a working paper*, London: DETR.

DETR (1999a) *A Better Quality of Life: the UK Sustainable Development Strategy*, London: DETR.

DETR (1999a) *Draft Guidance to the Director General of Water Services under Section 4*, London: DETR.

DETR (1999b) *Quality of Life Counts: Indicators for Sustainable Development*, Department for the Environment Transport and the Regions, London.

DETR (1999c) *Fuel Poverty: the New HEES – a Programme for Warmer, Healthier Homes*, London: DETR.

DETR (1999d) *Water Industry Act 1999: Consultation on Regulations*, London: DETR.

Devall, B. (1990) *Simple in Means, Rich in Ends*, London: Green Print.

Diefenbacher, H. (1994) 'The index of sustainable economic welfare in Germany', in Cobb and Cobb (eds), *The Green National Product*, Lanham: University of Americas Press.

DIW (1994) 'Ecological Tax Reform Even If Germany Has To Go It Alone', German Institute for Economic Research, *Economic Bulletin*, 37, Aldershot: Gower.

Dobson, A. (1990) *Green Political Thought*, London: Unwin Hyman.

Dobson, A. (1996) 'Representative Democracy and the Environment', in Lafferty, W. and Meadowcroft, J. (eds), *Democracy and the Environment*, Aldershot: Edward Elgar.

Dobson, A. (1998) *Justice and the Environment*, Oxford: Oxford University Press.

Dobson, A. (ed.) (1999) *Fairness and Futurity*, Oxford: Oxford University Press.

Dobson, A. (2000a) *Green Political Thought*, 3rd edition, London: Routledge.

Dobson, A. (2000b) 'Political theory and the environment: the grey and the Green (and the in-between)', in N. O'Sullivan (ed.), *Political Theory in Transition*, London: Routledge.

Doherty, B. and De Geus, M. (eds) (1996) *Democracy and Green Political Thought*, London: Routledge.

Dryzek, J. (1997) *The Politics of the Earth*, Oxford: Oxford University Press.

DSS (1998) *A New Contract for Welfare*, London: HMSO.

DSS (1999) *Opportunity for All: Tackling Poverty and Social Exclusion*, Cmnd 4445, London: HMSO.

DTI (2000) *New and Renewable Energy: Prospects for the 21st Century, Conclusions in Response to the Public Consultation*, London: DTI.

Dumas, C. (2000) 'How France wins the race for growth', *Guardian*, Monday, 28 February.

Durning, A. (1992) *How Much Is Enough?* , New York: W. W. Norton.

Durning, A. T. and Bauman, Y. (1998) *Tax Shift: How to Help the Economy, Improve the Environment, and Get the Tax Man Off Our Backs*, Northwest Environment Watch, NEW Report No. 7 (www.northwestwatch.org).

Eagleton, T. (1991) *Ideology: an Introduction*, London: Verso.

Eatwell, R. (1995) *Fascism: a History*, London: Chatto & Windus.

Eckersley, R. (1992) *Environmentalism and Political Theory*, London: UCL Press.

ECOTEC (1998) *Third System and Employment: Evaluation Inception Report*, Birmingham: ECOTEC.

Ehrenfeld, D. (1978) *The Arrogance of Humanism*, Oxford: Oxford University Press.

Ekins, P. (ed.) (1986) *The Living Economy*, London: Routledge.

Elgin, D. (1993) *Voluntary Simplicity*, New York: William Morrow.

Ellison N. (1999) 'Beyond Universalism and Particularism: Rethinking Contemporary Welfare Theory', *Critical Social Policy*, 19(1).

ENDS (1999a) *Environmental Data Services Report 291*, April 1999, London.

ENDS (1999b) *Environmental Data Services Report 288*, January 1999, London.

ENDS (1999c) *Environmental Data Services Report 289*, February 1999, London.

Environment Agency (2000) *Creating an Environmental Vision*, Bristol: Environment Agency.

Enzensberger, H. (1988) *Dreamers of the Absolute*, London: Radius.

Ernst, J. (1994) *Whose Utility? The Social Impact of Public Utility Privatization and Regulation in Britain*, Buckingham: Open University Press.

Esping-Andersen, G. (1999) *Social Foundations of Post-Industrial Economies*, Cambridge: Cambridge University Press.

Etzioni, A. (1995) *The Politics of Community*, London: Fontana.

European Commission (1996) *Social and Economic Inclusion Through Regional Development: the Community Economic Development Priority in ESF Programmes in Great Britain*, Brussels: European Commission.

European Commission (1997) *Towards an Urban Agenda in the European Union*, Communication from the European Commission COM (97)197.

European Commission (1998) *The Era of Tailor-made Jobs: Second Report on Local Development and Employment Initiatives*, Brussels: European Commission.

Faber, D. (1998) *The Struggle for Ecological Democracy*, New York: Guilford Press.

Fabian Society (2000) *Paying for Progress: a New Politics of Tax for Public Spending*, London: Fabian Society.

Ferris, J. (1993) 'Ecological Versus Social Rationality: Can There Be Green Social Policies?', in Dobson, A. and Lucardie, P. (eds), *The Politics of Nature*, London: Routledge.

Fitzpatrick, T. (1996) 'Postmodernism, Welfare and Radical Politics', *Journal of Social Policy*, 25(3).

Fitzpatrick, T. (1998) 'The Implications of Ecological Thought for Social Welfare', *Critical Social Policy*, 18(1).

Fitzpatrick, T. (1999a) *Freedom and Security: an Introduction to the Basic Income Debate*, London: Macmillan.

Fitzpatrick, T. (1999b) 'New Welfare Associations: an alternative model of well-being', in Jordan, T. and Lent, A. (eds), *Storming the Millennium*, London: Lawrence & Wishart.

Fitzpatrick, T. (2001a) 'Making Welfare for Future Generations', in Cahill, M. and Fitzpatrick, T. (eds), *Environmental Issues and Social Welfare*, special edition of *Social Policy & Administration*.

Fitzpatrick, T. (2001b) 'Dis/counting the Future', in Sykes, R., Ellison, N. and Bochel, C. (eds), *Social Policy Review 13*, Bristol: Policy Press.

Fitzpatrick, T. (2001c) 'New Agendas for Social Policy and Criminology', *Social Policy & Administration*, 35(2).

Fitzpatrick, T. (2002) 'The Two Paradoxes of Welfare Democracy' *International Journal of Social Welfare* 11(2).

Fitzpatrick, T. with Caldwell, C. (2001) 'Towards a Theory of Ecosocial Welfare: Radical Reformism and Local Currency Schemes', *Environmental Politics*, 10(2).

Fordham, G. (1995) *Made to Last: Creating Sustainable Neighbourhood and Estate Regeneration*, York: Joseph Rowntree Foundation.

Foster, J. (2000) *Marx's Ecology*, New York: Monthly Review Press.

Foucault, M. (1975) *Discipline and Punish*, London: Tavistock.

Foucault, M. (1977) *The History of Sexuality: Volume 1*, London: Allen Lane.

Foucault, M. (1984) *The Foucault Reader*, edited by Paul Rabinow, Harmondsworth: Penguin.

Frankel, B. (1987) *The Post-Industrial Utopians*, Cambridge: Polity Press.

Freeden, M. (1994) 'Political Concepts and Ideological Morphology', *The Journal of Political Philosophy* 2(2).

Freeden, M. (1996) *Ideologies and Political Theory*, Oxford: Oxford University Press.

Freeden, M. (1998) 'Is Nationalism a Distinct Ideology?', *Political Studies*, 46.

Fromm, E. (1976) *To Have or to Be?* London: Jonathon Cape.

Fukuyama, F. (1989) 'The End of History?', *The National Interest*, 16.

Fukuyama, F. (1992) *The End of History and the Last Man*, London: Hamish Hamilton.

Garland, D. (2001) *The Culture of Control*, Oxford: Oxford University Press.

George, V. and Wilding, P. (1994) *Welfare and Ideology*, Hemel Hempstead: Harvester Wheatsheaf.

Gershuny, J. (2000) *Changing Times*, Oxford: Oxford University Press.

Giddens, T. (1994) *Beyond Left and Right*, Cambridge: Polity.

Giddens, T. (1998) *The Third Way*, Cambridge: Polity.

Gilbert, B. B. (1966) *The Evolution of National Insurance in Great Britain*, London: Michael Joseph.

Girling, J. (1998) *France: Political and Social Change*, London: Routledge.

Goldsmith, E. (1996) *The Way: an Ecological World-view*, rev. edn, Totnes: Themis.

Goodin, R. (1992) *Green Political Theory*, Cambridge: Polity Press.

Gorz, A. (1989) *Critique of Economic Reason*, London: Verso.

Gorz, A. (1992) 'On the Difference Between Society and Community and Why Basic Income Cannot by Itself Confer Full Membership of Either', in Van Parijs, P. (ed.), *Arguing for Basic Income*, London: Verso.

Gorz, A. (1994) *Capitalism, Socialism, Ecology*, London: Verso.

Gorz, A. (1999) *Reclaiming Work: Beyond the Wage-Based Society*, Cambridge: Polity Press.

Gramsci, A. (1971) *Selections from Prison Notebooks*, London: Lawrence & Wishart.

Gray, J. (1987) 'Hayek as a Conservative', in Scruton, R. (ed.), *Conservative Thinkers*, London: Claridge Press.

Gray, J. (1993) *Beyond the New Right*, London: Routledge.

Gray, J. (1998) *False Dawn*, London: Granta.

Gregg, P., Johnson, P. and Reed, H. (1999) *Entering Work and the British Tax and Benefit System*, London: Institute for Fiscal Studies.

Groot, L. and van Iperen, F. (1999) 'Integraal Milieubeleid en het Groene Basisinkomen', *Milieu: Tijdschrift voor Milieuvraagstukken*, 3.

Guardian (1998) 'Raid the tax havens', 26 September.

Guenno, G. and Tiezzi, S. (1996) 'An Index of Sustainable Economic Welfare for Italy', FEEM Newsletter, 2.

Gyford, J. (1985) *The Politics of Local Socialism*, London: Allen & Unwin.

Habermas, J. (1975) *Legitimation Crisis*, London: Hutchinson.

Hajer, M. (1995) *The Politics of Environmental Discourse*, Oxford: Oxford University Press.

Hamilton, C. (1999) 'The Genuine Progress Indicator: methodological developments and results from Australia', *Ecological Economics*, 30(1).

Hamilton, M. B. (1987) 'The Elements of the Concept of Ideology', *Political Studies*, 35.

Hamond, M. (1997) *Tax Waste, Not Work: How Changing What We Tax Can Lead to a Stronger Economy and a Cleaner Environment*, Redefining Progress, (www.rprogress.org).

Harris, J. (1994) *Private Lives, Public Spirit: Britain 1870–1914*, Harmondsworth, Middx: Penguin.

Harrison, A. (1989) 'Introducing Natural Capital: the System of National Accounts', in Y. J. Ahmad, S. El Serafy and E. Lutz (eds), *Environmental Accounting for Sustainable Development*, Washington, DC: World Bank.

Haughton, G. (1998) 'Principles and practice of community economic development', *Regional Studies*, 32(9).

Hayward, T. (1994) *Ecological Thought: an Introduction*, Cambridge: Polity Press.

Hayward, T. (1997) *Political Theory and Ecological Values*, Cambridge: Polity.

Herber, L. (1963) *Our Synthetic Environment*, London: Jonathan Cape.

Herbert, A. and Kempson, E. (1995) *Water Debt and Disconnection*, London: Policy Studies Institute.

Herrington, P. (1996) *Climate Change and the Demand for Water. Department of the Environment*, London: HMSO.

Hicks, J. (1939) *Value and Capital*, Oxford: Oxford University Press.

Hills, B., Huby, M. and Kenway, P. (1997) *Fair and Sustainable: Paying for Water*, London: New Policy Institute.

Hills, J. (1998) *Thatcherism, New Labour and the Welfare State*, London: Centre for the Analysis of Social Exclusion Paper 13, London School of Economics.

Hindess, B. (1996) *Concepts of Power*, Oxford: Blackwell.

Hines, C. (2000) *Localization: a Global Manifesto*, London: Earthscan.

Hirsch, F. (1977) *Social Limits to Growth*, London: Routledge.

Hirst, P. (1994) *Associative Democracy*, Cambridge: Polity.

HM Treasury (1997) *Employment Opportunity in a Changing Labour Market*, London: HM Treasury.

HM Treasury (1998) *The Modernisation of Britain's Tax and Benefit System: the Working Families Tax Credit and Work Incentives*, London: HM Treasury.

Home Office (1999) *Community Self-help: Report of Policy Action Team no. 9*, London: Home Office.

Hoogendijk, W. (1991) *The Economic Revolution*, London: Merlin Press.

House of Commons Environmental Audit Committee (1999) *7th Report: Energy Efficiency*, London: The Stationery Office.

House of Commons Environment Select Committee (1996) *First Report: Water Conservation and Supply*, London: HMSO.

Huber, J. and Robertson, J. (2000) *Creating New Money: a Monetary Reform for the Information Age*, London: New Economics Foundation.

Huby, M. (1998) *Social Policy and the Environment*, Milton Keynes: Open University Press.

Huby, M. (2001) 'The Sustainable Use of Resources: The Global Picture', in M. Cahill and T. Fitzpatrick (eds), *Environmental Issues and Social Welfare*, special edition of *Social Policy & Administration*.

Huby, M. & Anthony, K. (1997) 'Regional Inequalities in Paying for Water', *Policy Studies*, 18.

Hulsberg, W. (1985) 'The Greens at the Crossroads', *New Left Review*, 152.

Hutton, W. (1996) *The State We're In*, London: Vintage.

Illich, I. (1977) *Disabling Professions*, London: Boyars.

International Union for Land Value Taxation (2000) 'Financing for Development', paper submitted to the United Nations (www.earthrights.net/docs/ fin4devt.html).

Irvine, A. S. and Ponton, A. (1988) *A Green Manifesto*, London: Macdonald Optima.

Islam, F. (2000) 'Vive la France – et la nouvelle economie', *Observer*, Sunday, 16 July.

Jackson, T. and Marks, N. (1994) *Index of Sustainable Economic Welfare*, Stockholm: Stockholm Environment Institute.

Jackson, T. and Marks, N. (1999) 'Consumption, Sustainable Welfare and Human Needs – with reference to UK expenditure patterns between 1954 and 1994', *Ecological Economics*, 28.

Jackson, T, and Stymne, S. (1996) *Sustainable Economic Welfare in Sweden – a pilot index 1950–1992*, Stockholm: Stockholm Environment Institute.

Jackson, T, Marks, N., Ralls, J. and Stymne, S. (1997) *Sustainable Economic Welfare in the UK: 1950–1996*, Centre for Environmental Strategy, London: New Economics Foundation.

Jacobs, M. (1996) *The Politics of the Real World*, London: Earthscan.

Jacobs, M. (1997) 'The Quality of Life: social goods and the politics of consumption', in M. Jacobs (ed.), *Greening the Millennium? The New Politics of the Environment*, Oxford: Blackwell.

Jacobs, M. (1999) *Environmental Modernisation*, London: Fabian Society.

Jessop, B. (1994) 'The Transition to Post-Fordism and Schumpeterian Workfare State', in R. Burrows and B. Loader (eds), *Towards a Post-Fordist Welfare State?*, London: Routledge.

Jessop, B. (2002) *The Future of the Capitalist State*, Cambridge:Polity Press.

Johnson, W. (1973) 'The Guaranteed Income as an Environmental Measure', in Daly, H. (ed.), *Toward a Steady-State Economy*, San Francisco: W. H. Freeman.

Jordan, B. (1992) 'Basic Income and the Common Good', in Van Parijs, P. (ed.) *Arguing for Basic Income*, London: Verso.

Jordan, B. (1998) *The New Politics of Welfare: Social Justice in a Global Context*, London: Sage.

Jordan, B., Agulnik, P., Burbidge, D. and Duffin, S. (2000), 'Stumbling towards a Basic Income – direction for tax and benefit reform', Citizen's Income Study Centre, London.

Jordan, B., James, S., Kay, H. and Redley, M. (1992) *Trapped in Poverty*, London: Routledge.

Kemp, P. and Wall, D. (1990) *A Green Manifesto for the 1990s*, Harmondsworth, Middx: Penguin.

Kempson, E. (1996) *Life on a Low Income*, York: Joseph Rowntree Foundation.

Klein, N. (2000) *No Logo*, London: HarperCollins.

Kraemer, S. (1997) *The Politics of Attachment: Towards a Secure Society*, London: Free Association Books.

Kuznets, S. (1971) *Economic Growth of Nations: Total Output and Production Structure*, Cambridge, Mass.: Harvard University Press.

Laborde, C. (1999) 'French Politics 1981–97: Stability and Malaise', in M. Cook and G. Davie (eds) *Modern France: Society in Transition*, London: Routledge.

Lambert, J. (1997) *No Change? No Chance!*, London: Jon Carpenter.

Larrain, J. (1979) *The Concept of Ideology*, London: Hutchinson.

Layfield Committee (1976) *Local Government Finance: the Report of the Committee of Inquiry*, London: HMSO.

Lee, K. (1993) 'To De-Industrialize: is it so Irrational?', in A. Dobson and P. Lucardie (eds) *The Politics of Nature: Explorations in Green Political Theory*, London: Routledge.

Lerner, S. (1994) 'The Future of Work in North America: Good Jobs, Bad Jobs, Beyond Jobs', *Futures*, 26(2).

Levett, R. (2001) 'Eco-efficiency of Quality of Life: presentation to a Foreign and Commonwealth Office environmental training workshop', CAG Consultants, London.

Lipietz, A. (1992) *Towards a New Economic Order*, Cambridge: Polity Press.

Little, A. (1996) *The Political Thought of Andre Gorz*, London: Routledge.

Little, A. (1998) *Post-Industrial Socialism: Towards a New Politics of Welfare*, London: Routledge.

Little, A. (2000) 'Environmental and eco-social rationality: challenges for political economy in late modernity', *New Political Economy*, 5(1).

Locke, J. (1960) *Second Treatise on Government*, Cambridge: Cambridge University Press.

Lorendahl, B. (1997) 'Integrating Public and Co-operative/Social Economy: towards a new Swedish model', *Annals of Public and Co-operative Economics*, 68(3).

Loughlin, M. (ed.) (1985) *Half a Century of Municipal Decline 1935–1985*, London: Allen & Unwin.

Macfarlane, R. (1996) *Unshackling the Poor: a Complementary Approach to Local Economic Development*, York: Joseph Rowntree Foundation.

Macintyre, A. (1982) *After Virtue*, London: Duckworth.

Markou, E. and Waddams Price, C. (1997) *Effects of UK Utility Reform: Source and Distribution*, IPPR Monograph, London: Institute for Public Policy Research.

Marx, K. (1977) *Selected Writings*, edited by D. McLellan, Oxford: Oxford University Press.

Maslow, A. (1954) *Motivation and Personality*, New York: Harper & Row.

Max Neef, M. (1991) *Human-Scale Development – Conception, Application and Further Reflection*, London: Apex Press.

Max Neef, M. (1995) 'Economic Growth and Quality of Life – a threshold hypothesis', *Ecological Economics*, 15.

Mayer, M. and Katz, S. (1985) 'Gimme Shelter: self-help housing struggles within and against the state in New York City and West Berlin', *International Journal of Urban and Regional Research*, 9(1).

McBriar, A. M. (1989) *An Edwardian Mixed Doubles*, Oxford: Oxford University Press.

McGinnis, M. (ed.) (1999) *Bioregionalism*, London: Routledge.

McHallam, A. (1991) *The New Authoritarians: Reflections on the Greens*, London: Alliance.

McKibben, B. (1990) *The End of Nature*, Harmondsworth, Middx: Penguin.

McLaren, D., Bullock, S. and Yousef, N. (1997) *Tomorrow's World*, London: Earthscan.

McLaughlin, A. (1994) *Regarding Nature: Industrialism and Deep Ecology*, New York: State University of New York Press.

McLellan, D. (1995) *Ideology*, Milton Keynes: Open University Press.

Meade, J. (1993) *Liberty, Equality and Efficiency*, London: Macmillan.

Meadows, D. H., Meadows, D. L., Randers, J. and Behrens, W. H. (1972) *The Limits to Growth*, London: Pan Books.

Mellor, M. (1992) *Breaking the Boundaries*, London: Virago.

Mestrovic, S. (1997) *Post-Emotional Society*, London: Sage.

Middleton, J. and Saunders, P. (1997) 'Paying for Water', *Journal of Public Health Medicine*, 19(1).

Midgley, M. (1983) *Animals and Why They Matter*, Harmondsworth, Middx: Penguin.

Mill, J. S. (1900) *Principles of Political Economy*, London: Longmans, Green & Co.

Miller, D. (1984) *Anarchism*, London: Dent.
Milner, S. and Mouriaux, R. (1997) 'France', in H. Compston (ed.), *The New Politics of Unemployment: Radical Policy Initiatives in Western Europe*, London: Routledge.
Mol, A. and Sonnenfeld, D. (2000) *Ecological Modernisation Around the World*, London: Frank Cass.
Morris, L. (1994) 'Informal Aspects of Social Divisions', *International Journal of Urban and Regional Research*, 18(1).
Morris, W. (1890) *News from Nowhere*, London: Longmans.
Morris, W. and Belfort Bax, E. (1893) *Socialism: Its Growth and Outcome*, London: Sonnenschein.
Myers, D. and Diener, E. (1996) 'The Pursuit of Happiness', *Scientific American*, 274.
Myers, N. (1998) *Perverse Subsidies: Tax $s Undercutting Our Economies and Environments Alike*, Winnipeg: International Institute for Sustainable Development.
Naess, A. (1973) 'The Sallow and the Deep, Long-Range Ecology Movement: A Summary', *Inquiry*, 16.
Naess, A. (1989) *Ecology, Community and Lifestyle*, Cambridge: Cambridge University Press.
Naess, A. (1995) 'Deep Ecology and Lifestyle', in G. Sessions (ed.), *Deep Ecology for the 21st Century*, Boston and London: Shambala.
National Consumer Council (1994) *Water Price Controls: Key Consumer Concerns*, London: National Consumer Council.
Nordhaus, W. (1992) 'Is growth sustainable?, Reflections on the Concept of Sustainable Economic Growth', paper for the International Economic Association, Varenna, Italy, October 1992.
Nordhaus, W. and Tobin, J. (1972) 'Is Growth Obsolete?' in *Economic Growth*, Fiftieth Anniversary Colloquium V, National Bureau of Economic Research, New York.
North, P. (1998) 'Exploring the Politics of Social Movements through "Sociological Intervention": a case study of Local Exchange Trading Schemes', *The Sociological Review*, 46(3).
North, P. (1999) 'Explorations in Heterotopia: Local Exchange Trading Schemes (LETS) and the micro-politics of money and livelihood', *Environment and Planning D: Society and Space*, 17(1).
Norton, B. (1984) 'Environmental Ethics and Weak Anthropocentrism', *Environmental Ethics*, 6(2).
Norton, B. (1987) *Why Preserve Natural Variety?*, Princeton: Princeton University Press.
Norton, B. (1991) *Toward Unity Among Environmentalists*, Oxford: Oxford University Press.
Nozick, R. (1974) *Anarchy, State and Utopia*, Oxford: Blackwell.
O'Connor, J. (1973) *The Fiscal Crisis of the State*, New York: St Martin's Press.
O'Connor, J. (1998) *Natural Causes: Essays in Ecological Marxism*, New York: Guilford Press.
O'Connor, M. (1994) *Is Capitalism Sustainable?*, New York: Guilford Press.
O'Neill, J. (1993) *Ecology, Policy, and Politics*, London: Routledge.
OECD (1996) *Reconciling Economy and Society: Towards a Plural Economy*, Paris: OECD.

Oegema, T. and Rosenberg, D. (1995) 'A Pilot ISEW for the Netherlands 1950-1992', Instituut vor Milieu en Systeemanalyse, Amsterdam.

Offe, C. (1984) *Contradictions of the Welfare State*, London: Hutchinson.

Offe, C. (1985) *Disorganised Capitalism*, Cambridge: Polity Press.

Offe, C. (1993) 'A Non-Productivist Design for Social Policies', in Coenan, H. and Leisnik, P. (eds), *Work and Citizenship in the New Europe*, Aldershot: Edward Elgar.

Offe, C. (1996) *Modernity and the State*, Cambridge: Polity Press.

Offe, C. and Heinze, R. (1992) *Beyond Employment*, Cambridge: Polity Press.

Offer, A. (1997) 'Between the Gift and the Market: the economy of regard', *Economic History Review*, 2.

OFWAT (1996) *Annual Report of the OFWAT National Customer Council, 1995–96*, Birmingham: Office of Water Services.

OFWAT (1999a) 'News Release, 17 June 1999', Birmingham: Office of Water Services.

OFWAT (1999b) *1999 Periodic Review. Final Determinations: Future Water and Sewerage Charges, 2000–05*, Birmingham: Office of Water Services.

OFWAT (1999c) *1998–1999 Report on Leakage and Water Efficiency*, Birmingham: Office of Water Services.

OFWAT (2000a) *Tariff Structure and Charges: 2000–01 Report*, Birmingham: Office of Water Services.

OFWAT (2000b) *Representing Water Customers 1999–2000. Annual Report of the Ofwat National Customer Council and the Ten Regional Customer Service Committees*, Birmingham: Office of Water Services.

OFWAT (2000c) *Approval of Companies' Charges Schemes 2001–2002 (A Consultation Paper by the Director General of Water Services)*, Birmingham: Office of Water Services.

Ophuls, W. and Boyan, S. (1992) *Ecology and the Politics of Scarcity Revisited*, New York: W. H. Freeman.

Oppenheim, C. (1998) 'Welfare to Work: taxes and benefits', in J. McCormick and C. Oppenheim (eds), *Welfare in Working Order*, London: IPPR.

Oswald, A. (1997) 'Happiness and Economic Performance', *Economic Journal*, 107.

Pacione, M. (1997a) 'Local Exchange Trading Systems as a Response to the Globalisation of Capitalism', *Urban Studies*, 34.

Pacione, M. (1997b) 'Local Exchange Trading Systems: a rural response to the globalisation of capitalism?', *Journal of Rural Studies*, 13(4).

Parsons, H. (1977) *Marx and Engels on Ecology*, London: Greenwood.

Pearce, F. (1991) *Green Warriors*, London: Bodley Head.

Pepper, D. (1993) *Eco-Socialism*, London: Routledge.

Perkin, H. (1989) *The Rise of Professional Society*, London: Routledge.

Pestoff, V. A. (1996) 'Work Environment and Social Enterprises in Sweden', paper presented to the European Conference on Labour Markets, Unemployment and Co-ops in Budapest, 27–28 October.

Philpott, J. (ed.) (1997) *Working for Full Employment*, London: Routledge.

Pierson, C. (1998) *Beyond the Welfare State?*, 2nd edn, Cambridge: Polity Press.

Polanyi, K. (1944) *The Great Transformation*, Boston, Mass.: The Beacon Press.

Porritt, J. (1984) *Seeing Green*, Oxford: Blackwell.

Porritt, J. (2000) *Playing Safe: Science and the Environment*, London: Thames & Hudson.
Porritt, J. and Winner, D. (1988) *The Coming of the Greens*, London: Fontana.
Postman, N. (1987) *Amusing Ourselves to Death*, London: Methuen.
Powell, M. (ed.) (1999) *New Labour, New Welfare State?*, Bristol: The Policy Press.
Powell, R. (1989) 'Toward Ecological Security', *Social Alternatives*, 9(1).
Putnam, R. (2000) *Bowling Alone*, New York: Simon & Schuster.
Radcliffe, J. (2000) *Green Politics: Dictatorship or Democracy?*, London: Macmillan.
Rawls, J. (1972) *A Theory of Justice*, Oxford: Oxford University Press.
Rawls, J. (1985) 'Justice as Fairness: Political not Metaphysical', *Philosophy and Public Affairs*, 14(2).
Redclift, M. (1996) *Wasted: Counting the Costs of Global Consumption*, London: Earthscan.
Reeves, R. (2001) *Happy Mondays*, Harlow: Pearson Education.
Regan, T. (1982) *All that Dwell Therein: Animal Rights and Environmental Ethics*, Berkeley: University of California Press.
Reid, D. (1995) *Sustainable Development: an Introductory Guide*, London: Earthscan.
Renooy, P. (1990) *The Informal Economy: Meaning, Measurement and Social Significance*, Amsterdam: Netherlands Geographical Studies no.115.
Rifkin, J. (1995) *The End of Work*, New York: G.P. Putnam.
Robbins, D. (1994) *Social Europe. Towards a Europe of Solidarity: Combating Social Exclusion*, Brussels: European Commission.
Robertson, J. (1989) *Future Wealth: A New Economics for the 21st Century*, London: Cassell.
Robertson, J. (1994) *Benefits and Taxes: a Radical Strategy*, London: New Economics Foundation.
Robertson, J. (1998) *Transforming Economic Life: a Millennial Challenge*, Schumacher Briefing No. 1, Dartington: Schumacher Society/Green Books.
Robertson, J. (1999) *The New Economics of Sustainable Development: a Briefing For Policy Makers*: London: Kogan Page.
Robertson, J. (2000) *Financial and Monetary Policies for an Enabling State*, the Alternative Mansion House Speech, given at an event organised by the New Economics Foundation, 15 June.
Robinson, M. (1992) *The Greening of British Party Politics*, Manchester: Manchester University Press.
Robinson, P. (1998) 'Employment and Social Inclusion', in C. Oppenheim (ed.), *An Inclusive Society: Strategies for Tackling Poverty*, London: IPPR.
Rodger, J. J. (2000) *From a Welfare State to a Welfare Society*, Basingstoke: Macmillan
Roemer, J. (1994) *A Future for Socialism*, London: Verso.
Room, G. (ed.) (1995) *Beyond the Threshold: the Measurement and Analysis of Social Exclusion*, Bristol: Policy Press.
Rose, N. (1999a) *Governing the Soul*, 2nd edn, London: Free Association Books.
Rose, N. (1999b) *Powers of Freedom*, Cambridge: Cambridge University Press.
RCEP (2000) *Twenty-second Report. Energy – The Changing Climate*, Cm 4749, London: The Stationery Office.
Sagoff, M. (1988) *The Economy of the Earth*, Cambridge: Cambridge University Press.

Sale, K. (1985) *Dwellers in the Land: the Bioregional Vision*, San Francisco: Sierra Club.

Saunders, P. and Harris, C. (1994) *Privatization and Popular Capitalism*, Buckingham: Open University Press.

Saward, M. (1996) 'Must Democrats be Environmentalists?', in Doherty, B. and De Geus, M. (eds), *Democracy and Green Political Thought*, London: Routledge.

Scitovsky, T. (1976) *The Joyless Economy*, New York: Oxford University Press.

Sennett, R. (1998) *The Corrosion of Character*, London and New York: W. W. Norton & Company.

Seyfang, G. (1998) *Green Money from the Grassroots: Local Exchange Trading Schemes and Sustainable Development*, Submitted PhD thesis, Leeds Metropolitan University.

Shaw, G. B. (1889) *Fabian Essays in Socialism*, London: Fabian Society.

Social Exclusion Unit (1998) *Bringing Britain Together: a National Strategy for Neighbourhood Renewal*, London: Social Exclusion Unit.

Solomon, J. (1998) *To Drive or To Vote?*, London: Chartered Institute of Transport.

Spretnak, C. and Capra, F. (1985) *Green Politics*, London: Paladin.

Stavrakakis, Y. (1997) 'Green Ideology: a Discursive Reading', *Journal of Political Ideologies*, 2(3).

Stavrakakis, Y. (2000) 'On the Emergence of Green Ideology: the Dislocation Factor in Green Politics', in Howarth, D., Norval, A. and Stavrakakis, Y. (eds.), *Discourse Theory and Political Analysis*, Manchester: Manchester University Press.

Stockhammer, E., Hochreiter, H., Obermayr, B. and Stelner, K. (1997) 'The Index of Sustainable Economic Welfare (ISEW) as an Alternative to GDP Measuring Economic Welfare: the results of the Austrian (revised) ISEW calculation 1955 – 1992', *Ecological Economics*, 21(1).

Stoker, G. (1987) 'Decentralisation and Local Government', *Social Policy and Administration*, 21:2.

Stoker, G. (ed.) (2000) *The New Governance of Local Government*, London: Macmillan.

Stringer, E. T. (1996) *Action Research: a Handbook for Practitioners*, London: Sage.

Stymne, S. and Jackson, T. (2000) 'Intragenerational Equity and Sustainable Welfare', *Ecological Economics*, 33.

Thomas, J. J. (1992) *Informal Economic Activity*, Hemel Hempstead: Harvester Wheatsheaf.

Thompson, J. (1984) *Studies in the Theory of Ideology*, Oxford: Oxford University Press.

Tokar, B. (1992) *The Green Alternative*, San Pedro: R & E Miles.

Torfing, J. (1999) *New Theories of Discourse*, Oxford: Blackwell.

Torgerson, D. (1999) *The Promise of Green Politics*, Durham, NC: Duke University Press.

Trainer, F. E. (1998) 'The Significance of the Limits to Growth for the Discussion of Social Policy', *International Journal of Sociology and Social Policy*, 18(11/12).

Tylecote, A. (2000) 'France gets to Heart of Problem', *Guardian*, Monday 7 August.

Urry, J. (2000) *Sociology Beyond Societies: Mobilities for the Twenty-first Century*, London: Routledge.

Van Dijk, T. A. (1998) *Ideology: a Multidisciplinary Approach*, London: Sage.
Van Parijs, P. (1991) 'Basic Income: a Green strategy for the new Europe', in Parkin, S. (ed.), *Green Light on Europe*, London: Heretic Books.
Van Parijs, P. (ed.) (1992) *Arguing for Basic Income*, London: Verso.
Veblen, T. (1899) *The Theory of the Leisure Class*, London: Prometheus Books.
Vincent, A. (1993a) *Modern Political Ideologies*, Oxford: Blackwell.
Vincent, A. (1993b) 'The Character of Ecology', *Environmental Politics*, 2(2).
Wall, D. (1990) *Getting There: Steps to a Green Society*, London: Green Print.
Weitzman, M. (1976) 'On the Welfare Significance of the National Product in a Dynamic Economy', *Quarterly Journal of Economics*, 90.
Weizsacker, E. (1994) *Earth Politics*, London: Zed Books.
Weizsacker, E., Lovins, A. and Lovins, L.H. (1998) *Factor Four*, London: Earthscan.
Westerdahl, S. and Westlund, H. (1998) 'Social Economy and New Jobs: a summary of twenty case studies in European regions', *Annals of Public and Co-operative Economics*, 69(2).
White, L. (1967) 'The Historic Roots of Our Ecological Crisis', *Science*, 155.
Whitelegg, J. (1997) *Critical Mass: Transport, Environment and Society in the Twenty-First Century* London: Pluto Press.
Williams, C. C. and Windebank, J. (1999) *A Helping Hand: Harnessing Self-help to Combat Social Exclusion*, York: Joseph Rowntree Foundation.
Williams, C. C. and Windebank, J. (2000) 'Helping Each Other Out? Community exchange in deprived neighbourhoods', *Community Development Journal*, 35(2).
Williams, C. C. (1996a) 'Local Exchange and Trading Systems (LETS): a new form of work and credit for the poor and unemployed', *Environment and Planning A*, 28(8).
Williams, C. C. (1996b) 'The New Barter Economy: an appraisal of Local Exchange and Trading Systems (LETS)', *Journal of Public Policy*, 16(1).
Williams, C. C. (1996c) 'Informal Sector Responses to Unemployment: an evaluation of the potential of Local Exchange and Trading Systems (LETS)', *Work, Employment and Society*, 10(2).
Wilson, E. O. (1984) *Biophilia*, Cambridge, Mass.: Harvard University Press.
Wilson, R. (1998) 'Comment: citizen's involvement', in C. Oppenheim (ed.), *An Inclusive Society: Strategies for Tackling Poverty*, London: IPPR.
Windebank, J. and Williams, C. C. (1995) 'The Implications for the Informal Sector of European Integration', *European Spatial Research and Policy*, 2(1).
Wissenburg, M. (1997) 'A Taxonomy of Green Ideas', *Journal of Political Ideologies*, 2(1).
Wissenburg, M. (1998) *Green Liberalism*, London: UCL Press.
Worcester, R. (1998) 'More than Money', in I. Christie and L. Nash (eds), *The Good Life*, London: Demos.
Wright, R. (1995) 'The Evolution of Despair', *Time*, 28 August 1995.
Yeo, S. (1976) *Religion and Voluntary Organisations in Crisis*, London: Croom Helm.

Index